D0758110

ACID RAIN

BOOKS BY ROBERT H. BOYLE

The Hudson River: a Natural and Unnatural History
Malignant Neglect (with the Environmental Defense Fun
Bass
Stoneflies for the Angler (with Eric Leiser)
At the Top of Their Game

ACID RAIN

Robert H. Boyle
and R. Alexander Boyle

NICK LYONS BOOKS
SCHOCKEN BOOKS • NEW YORK

To Katya
with love and thanks

First published by Schocken Books 1983

10 9 8 7 6 5 4 3 83 84 85 86

Produced by Nick Lyons Books
212 Fifth Avenue
New York, NY 10010

Library of Congress Cataloging in Publication Data

Boyle, Robert H.
Acid rain.

Includes index.
1. Acid rain. I. Boyle, R. Alexander. II. Title.
TD196.A25B69 1983 363.7'394 82-21410

Manufactured in the United States of America

ISBN 0-8052-3854-9 (hardback)
0-8052-0746-5 (paperback)

Some of the material in this book originally appeared, in different form, in *Sports Illustrated*.

Acknowledgments

Much of the material in this book is based on our own reporting, while the remainder is drawn from the scientific literature and other published sources cited in the Bibliography.

We would like to express our thanks to Gilbert Rogin, Peter Carry, Bob Brown, Jerry Kirshenbaum, Gay Flood, Rose Mary Mechem and Robert Sullivan of *Sports Illustrated* (some of the material in the book first appeared in the magazine); Dr. Ernest W. Marshall of West Stockholm, New York; David M. Seymour of Santa Fe, New Mexico; Dr. Michael Oppenheimer and J. Andrew Hudis of the Environmental Defense Fund, New York City; Richard Ayres and David Hawkins of the Natural Resources Defense Council, Washington, D.C.; Adele

Hurley and Michael Perley of the Canadian Coalition on Acid Rain, Washington, D.C., and Toronto; F. Peter Fairchild of the Boston College Observatory, Weston, Massachusetts; Drs. Hubert Vogelmann and Richard Klein of the University of Vermont; Dr. Eville Gorham of the University of Minnesota; Dr. Ellis B. Cowling of North Carolina State University; Major John Robertson of the United States Military Academy, West Point; Ted Williams of Grafton, Massachusetts; Fred Johnson of the Pennsylvania Fish Commission, Harrisburg; Dr. Harold H. Harvey of the University of Toronto; Pamela K. McClelland of Trout Unlimited, Vienna, Virginia; Paul Hansen of the Izaak Walton League of America, Arlington, Virginia; Dr. Jay D. Hair of the National Wildlife Federation, Washington, D.C.; Peter Silverstein of New York City; Harriett S. Stubbs of West St. Paul, Minnesota; Dr. Dominick J. Pirone of Manhattan College; Lennart Borgstrom of North Caldwell, New Jersey, and Svängsta, Sweden; Ruth Eisenhower of Cold Spring, New York; and Nick Lyons and Peter Burford of New York City.

ROBERT H. BOYLE and R. ALEXANDER BOYLE

Contents

. . . the dripping of a poisoned rain
is all the burning sand receives.

PUSHKIN, "The Upas Tree"
(translated by Vladimir Nabokov)

Dogged with dew, dappled with dew
Are the groins of the braes that the brook treads through
Wiry heathpacks, flitches of fern,
And the beadbonny ash that sits over the burn.

What would the world be, once bereft
Of wet and wildness? Let them be left,
O let them be left, wildness and wet;
Long live the weeds and the wildness yet.

GERARD MANLEY HOPKINS,
"Iversnaid"

1

The Scope of the Problem

A chemical leprosy is eating into the face of North America and Europe. This chemical leprosy is commonly called acid rain. Scientists have generally used the term "acid precipitation"—the one we have used—because of the occurrence of acid snow, acid sleet, acid hail, acid fog, acid frost, and acid dew, as well as acid rain. Now, in increasing recognition of the devastating role that "dry" deposits of acid particles, aerosols, and gases can play in this appalling phenomenon, a new term is coming into widespread use: "acid deposition."

Acid deposition, acid precipitation, acid rain—whatever name you want to use—is caused by the emission of sulfur dioxide and nitrogen oxides from the combustion of fossil fuels. Natural sources, such as volcanoes and mud flats, emit sulfur dioxide to the atmosphere, but their contribution is small. "About ninety percent of the sulfur in the atmosphere of the northeastern United States comes from man-made sources," says Dr. George R. Hendrey, leader of the Environmental Sciences Group at the Brookhaven National Laboratory in Upton, New York. Aloft in the atmosphere, the sulfur dioxide and nitrogen oxides can be transformed into sulfuric acid and nitric acid, and air currents can carry them hundreds and sometimes thousands of miles from their source. When these acids fall to earth in whatever form, they can have a devastating and possibly irreversible impact on lands and waters that have low natural buffering capacity.

Acid precipitation has been called the single most important environmental threat to the United States and Canada, but it is important to bear in mind that acid precipitation is only one manifestation, albeit a very serious one, of the widespread pollution of the atmosphere. Man's activities have made the sky a sewer. Each year the global atmosphere is on the receiving end of 20 billion tons of carbon dioxide, 130 million tons of sulfur dioxide, 97 million tons of hydrocarbons, 53 million tons of nitrogen oxides, more than three million tons of arsenic, cadmium, lead, mercury, nickel, zinc, and other toxic metals, and a host of synthetic organic compounds ranging from polychlorinated biphenyls (PCBs) to toxaphene and other pesticides, a number of which may be capable of causing cancer, birth defects, or genetic change.

Pollutants are all over the place. Take lead. A 1981 report, *Atmosphere-Biosphere Interactions: Toward a Better*

Understanding of the Ecological Consequences of Fossil Fuel Combustion, by the National Research Council of the National Academy of Sciences notes, "Natural lead releases have now been exceeded by emissions from man's activities. While data are scarce, some have argued that the present body burden of lead in average Americans, and probably in other residents of industrial societies, is approaching chronically damaging levels." That's just lead. Cadmium and zinc are coming down at such a rate that scientific models suggest that both these metals will reach concentrations toxic to zooplankton in Lake Michigan sometime within the next thirty to eighty years. Zooplankton is the microscopic animal life at the base of the entire food web in the lake, and Lake Michigan is not some one-acre farm pond that could be done in easily. It happens to be the sixth largest lake in the world in both surface area and volume.

Then there are the problems posed by the interaction of one pollutant with another. In addition to contributing to acid precipitation, nitrogen oxides can react with hydrocarbons to produce ozone, a major air pollutant responsible in the United States for annual losses of from $2 billion to $4.5 billion worth of wheat, corn, soybeans, and peanuts. A wide range of interactions can occur with toxic metals. As Dr. F. Herbert Bormann of the School of Forestry and Environmental Studies at Yale University says, "It's a vicious web." The scientists who wrote the 1981 report of the National Research Council concluded that "It is the Committee's opinion, based on the evidence we have examined, that the picture is disturbing enough to merit prompt tightening of restrictions on atmospheric emissions from fossil fuels and other large sources. . . . Strong measures are necessary, if we are to prevent further degradation of natural ecosystems, which together support life on this planet."

Despite all this, the Reagan Administration is trying to reverse the environmental gains that were made in the 1960s and the 1970s. Instead of seeking to strengthen the Clean Air Act, which did not address the problem of acid precipitation when it was written in 1970, the Reagan Administration has been attempting to gut it. It is ironic that an administration that would have us believe, based on secondhand evidence, that Soviet chemical warfare has caused "yellow rain" to fall in Laos and Cambodia, questions the occurrence of pollution-caused acid rain here at home where the evidence is clear, compelling, and available on almost a daily basis.

Consider the following:

• Acid precipitation has killed fish and other aquatic life outright. In Scandinavia, which is downwind of pollution pumped into the skies of Western Europe, acid precipitation has destroyed fish life in 5,000 lakes in southwestern Sweden and in 1,500 lakes and seven Atlantic salmon rivers in southern Norway. In Canada, southwestern Sweden and in 1,500 lakes and seven Atlantic salmon rivers in southern Norway. In Canada, Ontario has lost fish populations in an estimated 1,200 lakes, and provincial authorities calculate that Ontario stands to lose the fish in 48,500 lakes within the next twenty years if acid precipitation continues at the present rate.

In New York, acid precipitation has rendered 212 lakes in the Adirondacks unfit for fish. It is also doing in lakes and ponds in the Hudson Highlands near New York City.

Acid precipitation is also falling on sensitive lands and waters in Massachusetts, Rhode Island, Vermont, New Hampshire, Maine, New Jersey, Pennsylvania, Maryland, and West Virginia. For example, acidification threatens 150 miles of West Virginia's total of 550 miles of native brook trout streams, and in Massachusetts acid precipitation is endangering the Quabbin Reservoir, which supplies the drinking water for more than two million people in the greater Boston area. In 1981 an official of the U.S. Environmental Protection Agency told Senator George Mitchell of Maine, who was concerned about what was happening to his state, that the Reagan Administration would not take any action on acid precipitation as long as it was confined to the Northeast. The administration should do its homework. Acid precipitation is hitting Southeastern states from Florida to Kentucky. Rain that falls on Raleigh, North Carolina, is sometimes more acid than white vinegar, and on its weather page *The Charlotte Observer* routinely reports on the acidity of the rain, a practice that should be adopted nationally by other newspapers and by TV and radio stations. Acid precipitation, often at levels associated with the onset of lake acidification in Scandinavia, is now falling on sensitive lands and waters in Michigan and Minnesota. Acid precipitation is also falling on Colorado, Idaho, New Mexico, Montana, California, and Washington. Dr. Thomas Crocker of the University of Wyoming calculates that acid precipitation is causing more than $250 million a year in damage to aquatic ecosystems throughout the United States.

• Acid precipitation can have damaging effects on human health. It can poison reservoirs and water-supply systems because it can mobilize (that is, put into circulation) mercury, cadmium and other potentially toxic metals (including aluminum which is under suspicion) in the bedrock and soils in a watershed. Acid drinking water can also corrode metals, such as lead and copper, used in water-supply pipes.

Human health can also be adversely affected in other ways. Sulfates and other fine particles transported in the atmosphere may cause cancer and cardiovascular diseases. An Jennie E. Bridge and F. Peter Fairchild have stated in the *Northeast Damage Report*, "The probability of dying from air-pollution-related diseases is twice as high in the Northeast than in other regions of the United States."

• Acid precipitation may pose a menace to crops and forests. Some of the crops known to be sensitive to sulfur dioxide include alfalfa, barley, cabbages, peas, rye, soybeans, spinach, and tobacco. Some of the sensitive trees include white ash, large tooth aspen, birch, elm, maple, and ponderosa pine.

• Acid precipitation is disfiguring buildings and monuments, including the Capitol in Washington. The *Northeast Damage Report* notes that the sandstone of the Essex Institute, built in 1850 in Salem, Massachusetts, has suffered extraordinary erosion between 1964 and the present. Indeed, if erosion during the first 115 years had occurred at the rate observed in the last fifteen years, there would be no stone on the building today.

Acidity is measured on the pH—literally the potential of Hydrogen—scale, which runs from zero to 14. Seven is neutral: the numbers above increasingly alkaline, the numbers below increasingly acidic. The pH scale is logarithmic, so that water with a pH of 5 is ten times more acidic than water with a pH of 6, and water with a pH of 4 is one hundred times more acidic than water with a pH of 6.

"Pure" water is slightly acidic. It has a pH of between 5.6 and 5.7 because the water molecules combine

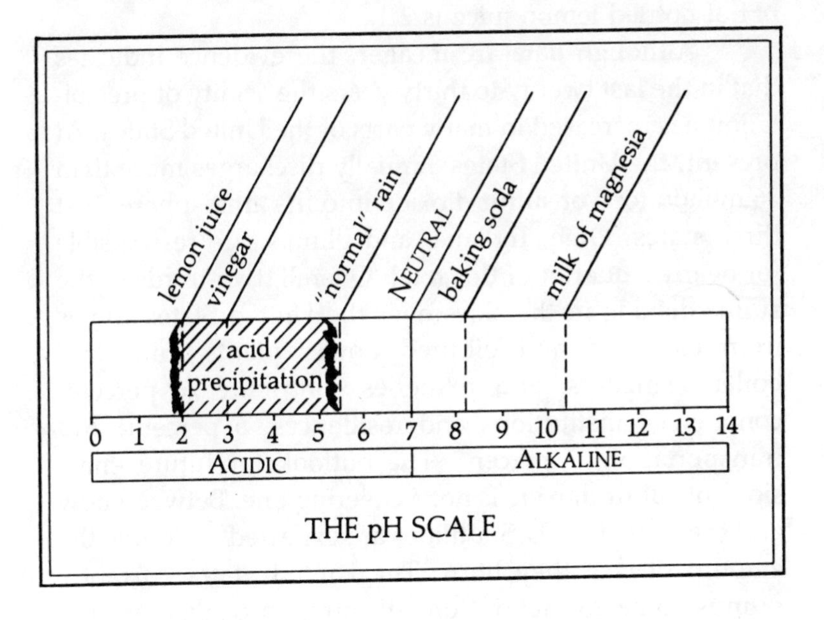

THE pH SCALE

with carbon dioxide naturally present in the atmosphere to form weak carbonic acid. Although scientists are not certain that rain was ever "pure"—for instance, a million years ago rain may have had pHs as high as 6 or 7 as the result of free-floating alkaline dust—they do know that trouble begins when the acidity of precipitation is intensified. At present, the median pH for precipitation in the Northeast is generally accepted to be 4.3. Although not "officially measured," the lowest recorded pH for a single storm anywhere in the world occurred in Wheeling, West Virginia, in the fall of 1978. According to officials of the U.S. EPA in Wheeling, pHs of less than 2 were discovered in the course of a three-day drizzle. That rain was 5,000 times more acid than normal. For comparison, the pH of bottled lemon juice is 2.1.

Although data are meager, the evidence indicates that in the last twenty to thirty years the acidity of precipitation has increased in many parts of the United States. At present, the United States annually discharges more than 26 million tons of sulfur dioxide into the atmosphere. Just three states, Ohio, Indiana, and Illinois, are responsible for nearly a quarter of this total. Overall, two-thirds of the sulfur dioxide in the skies over the United States comes from coal-fired and oil-fired power plants. Industrial boilers, smelters, and refineries contribute 26 percent; commercial institutions and residences, 5 percent; and transportation, 3 percent. The outlook for future emissions of sulfur dioxide is not a cheering one. Between now and the year 2000, U.S. utilities are expected to double the amount of coal they burn. The United States currently pumps some 23 million tons of nitrogen oxides into the atmosphere in the course of a year. Transportation sources account for 40 percent; power plants, 30 percent; industrial sources, 25 percent; and commercial institutions and residences, 5 percent. What makes these figures particularly disturbing is that nitrogen oxide emissions have *trebled* in the last thirty years.

After the Clean Air Act was passed in 1970, utilities often sought to meet local air standards by building very tall stacks. The EPA calculates that before 1970 there were only two stacks more than 500 feet high in the United States. Now there are 180 such stacks. Tall stacks can relieve local air pollution, but they increase acid precipitation in downwind areas. A tall stack can be likened to a missile silo, capable of hurling aerial garbage at target areas far away. In 1974 American Electric Power ran an advertisement boasting that it was the "pioneer" of tall stacks. The ad proclaimed that the tall stacks dispersed "gaseous emissions widely in the atmosphere so that ground level concentrations would not be harmful to human health or property." These emissions, American Electric Power claimed, were "dissipated high in the atmosphere, dispersed over a wide area, and come down finally in harmless traces." The ad derided "irresponsible environmentalists" who wanted strict controls over the emissions and accused them of "taking food from the mouths of the people to give [themselves] a better view of the mountain." It is now clear that those "harmless traces" can be lethal to many forms of life.

Although no one has so far actually traced acid precipitation falling on Lake X in Maine or Lake Y in Quebec to Power Plant Z in Ohio, satellite photographs show the linkage between local, regional, and global pollution. Dr. Walter A. Lyons is president of Mesomet, Inc., a Chicago research firm commissioned by the U.S. EPA to do a film called "Satellite Observations of Persistent Elevated Pollution Episodes," known as "PEPEs" or "Smog Blobs" in the trade. Dr. Lyons showed the film to the House Subcommittee on Oversight and Investigations of the Committee on Interstate and Foreign Commerce in February of 1980, and here is part of his narration, starting with a PEPE that occurred on one day in June 1975:

> Note the high-pressure system over the Great Lakes, producing a clockwise flow of air. Pollution is coming from the northeast through the Ohio Valley, penetrating as far west as central Oklahoma and Kansas, then streaming northward into Minnesota and Wisconsin.
>
> The notion that air simply moves from west to east is true only as the grossest generalization. Air can move in any direction it wants.
>
> You might also notice some other interesting characteristics. In Missouri there is a thin spot in the haze moving southwestward from St. Louis. This is apparently the footprint of a prior thunderstorm which washed out a large amount of the sulfate aerosols and presumably was involved in an acid rain event.
>
> You cannot see acid rain but you can see a signature, an indicator in the haze patterns where acid rain has occurred.

At the northern terminus of this airborne sewer you can see tremendous amounts of low-visibility, sulfate-laden air moving north into Minnesota. Note the tremendous thunderstorms in the northern part of the state, not far from the Boundary Waters Canoe Area. Most of the pollution which had come from points far to the south and east was being fed into those thunderstorms and possibly was causing high concentrations of sulfate and low pH values in the rainwater. What did not get into the rainstorms possibly moved across the border into Canada.

The effects of these events are very difficult to quantify. I would not presume to make any definitive statements about the effect of this complex of acid rain, low visibility, sulfate, and ozone. The point I want to make is that they all tend to occur in tandem. There is not a single problem. If you have one you have typically the other. Research done in central Minnesota has noticed [this situation] when we have large-scale intrusions of polluted air from the south, into Minnesota, apparently far removed from the effects of the Twin Cities.

A typical situation may show an initially clean high-pressure system drifting into the Ohio Valley region. The circulation around it picks up pollutants from various areas, not simply Ohio but many states. Ohio perhaps receives as much as it gives in this situation.

Here in this example Iowa and Minnesota are now subject to several days' intrusion of ozone near

> or above the federal standard, sulfate levels above twenty micrograms, and low visibility. Under these circumstances, if it occurs during particularly sensitive periods in a growth cycle of a soybean crop, there is evidence to suggest that there might be as much as a 15 to 20 percent yield reduction—might be. It is not definitive. If you are not worried about acid rain or sulfates, there may be other chemicals which are causing problems at the same time, such as photochemical oxidants. The ozone similarly transported may not be causing a beneficial result to agriculture.

Another series of photographs showed what Dr. Lyons called "a widespread pollution episode over the East Coast." He noted, "This tremendous cloud of pollutants, presumably mostly sulfate aerosols, was drifting not only offshore but more than a thousand miles eastward across the Atlantic, making us wonder if pollutants in significant quantities from the Ohio Valley could not reach Europe under favorable transport conditions."

The effect that acid precipitation has on a body of water depends on the nature of the rocks and soils in the watershed. A watershed with calcareous soils or rocks containing calcium carbonate, calcium, or lime can buffer acid precipitation in much the same way that an Alka-Seltzer or a Rolaids tablet neutralizes an acid stomach. Some parts of North America, such as the Dakotas, which have alkaline soils, have great buffering capacity, but there are other areas that have hard rock and/or infertile sandy soils, and these have minimal buffering capacity. Geologic outcroppings and anomalies can make for vast differences within an area. How much acid precipitation it takes to acidify a specific body of water depends on the acid-neutralizing capacity of that body of water, chemically measured as its total alkalinity. A body of water that has a total alkalinity of 100 microequivalents or less per liter is considered very susceptible to acidification. Similarly a body of water with a total alkalinity of 100 to 500 microequivalents per liter is considered potentially susceptible, and one with more than 500 microequivalents per liter is deemed not susceptible. Sometimes total alkalinity is also measured in parts per million. An alkalinity of 5 ppm is the same as an alkalinity of 100 microequivalents per liter.

Knowledge of the total alkalinity of a body of water and whether or not that alkalinity is decreasing is essential because the pH can be deceptive, dropping sharply only as the buffering capacity is finally destroyed. In experimental acidification of a small lake in Canada, 70 percent of the alkalinity was depleted before pH values dropped detectably below normal.

Snowfall can play a key role in acidification. Dr. Ernest W. Marshall, a geologist who studies snow and ice as though they were temporary rock formations, can track different storms through the Adirondacks weeks after they have occurred by digging into the snowpack and examining individual storm layers. To Marshall it is no coincidence that the Adirondack lakes that suffer most each spring lie on the western slope of the range. They receive the greatest amount of snow, and they lie directly downwind of polluted air masses coming from the Middle

West, the Ohio River Valley, and southern Canada. "The snow and ice store acids for three to four months," says Marshall, "and then when the spring melt comes, lakes and streams get one hell of a slug of acids. It's as though a pack-a-day smoker gave up cigarettes for four months and then tried to make up for what he had missed by smoking 120 packs in ten days."

High acid episodes during snowmelt are not unusual in lakes and streams that are otherwise normal, and large fish kills can result. We wonder what occurred in New York and adjacent states during the spring of 1982. One snowmelt occurred in March, and then on April 6 a blizzard deposited ten to fifteen inches of snow. For all the coverage given the blizzard, not one newspaper, wire service, or radio or TV station reported that the pH of the snow was 4.4. The snow melted rapidly in the next week, and instead of one acid pulse, the spring of 1982 had two. What happened to fish in lakes and streams? What happened to sensitive amphibians, such as the spring peepers, that breed and lay their eggs in shallow pools? Rachel Carson wrote of a silent spring, but acid precipitation with its capacity to destroy life at the start offers the possibility of a silent summer, a silent fall, and a silent winter as well.

No portion of the United States has been harder hit than the Adirondacks, ancient mountains with rock more than a billion years old. They are not part of the Appalachians but are a southward extension of the Canadian Shield, which includes the Laurentians. At one time the ancestral range towered as high as the Himalayas, but erosion, subsidence, and glaciation have worn the higher peaks down to their tough roots of anorthosite—a very durable rock, the nether parts of which are embedded, like giant molars, six miles deep in the earth. The soil cover, where present, is scanty and infertile.

It is ironic that the Adirondacks should be suffering the onslaughts of acid precipitation because no part of the United States is more protected. The Adirondack Park covers six million acres—almost 10,000 square miles—and it is the biggest park in the country. It is almost three times the size of Yellowstone; it is, in fact, bigger than the State of Massachusetts. New York State owns 2.3 million acres of the park outright; this land is locked up as much as any wilderness can be as a Forest Preserve. As the result of public protest against the devastating logging practices in the nineteenth century, a unique clause in the state constitution decrees that the Forest Preserve lands "shall be forever kept as wild forest lands." Except for the maintenance of trails and campsites, no trees on state lands can be destroyed, removed, or sold without amending the constitution, and that would require passage by two successive legislatures and statewide approval by the voters. This protection has been in force since 1895, and in all that time no constitutional convention or session of the legislature has ever adopted any amendment to repeal the "forever wild" clause.

The remaining 3.7 million acres within the park boundary, popularly known as the Blue Line, are privately owned. These private lands are given over to estates, villages, resorts, hunting and fishing camps, lumber camps, mines, and farms. The private and forever wild lands often intermingle like the pieces of a jigsaw puzzle, but the use of any private property must conform to the Adirondack Park and Land Use Development Act adopted as state law in 1973. Administered by the independent, bipartisan Adirondack Park Agency, the Adirondack Park is the largest single area in the United States and perhaps the world under one comprehensive land-use plan. It is a glorious region. About 100,000 acres of state land have never been cut at all, and this primeval forest imparts to the visitor a sense of timelessness that is extraordinarily rare in this age of future schlock. As William Chapman White wrote in *Adirondack Country*, "As a man tramps the woods to the lake he knows he will find pines and lilies, blue heron and golden shiners, shadows on the rocks and the glint of light on the wavelets, just as they were in the summer of 1354, as they will be in 2054 and beyond. He can stand on a rock by the shore and be in a past he could not have known, in a future he will never see. He can be part of time that was and time yet to come."

That poetic passage was written in 1966 before anyone in the Adirondacks was aware of acid precipitation. Now the shiners are gone from some lakes and the blue heron is disappearing. "It's insidious," said C. V. Bowes, Jr., the owner of Covewood Lodge, a resort on Big Moose Lake in the western Adirondacks. It is late in the afternoon in May 1981, and Bowes is standing in the living room of his house looking out on the acid waters of Big Moose Lake. It was on Big Moose Lake in 1906 that Chester Gillette drowned his mistress, Grace Brown, thereby providing Theodore Dreiser with the plot for his novel, *An*

American Tragedy, a prophetic title in view of what has happened to the lake in recent years. The still pristine-appearing lake is now so acid that swimmers sometimes emerge with bloodshot eyes, and except for the few odd fish that hover about spring holes in the bottom, the trout are gone.

"I can remember how good the fishing was," Bowes said. "Then thirty years ago, it slowly started to tail off. It's not something that happened overnight. It built up slowly. As the fishing got worse, the guides began moving away for other jobs. First the state blamed it on the big blowdown of 1950, when we lost 75 percent of our coniferous trees in the aftermath of a hurricane. The state said that the downed trees made the water poorer for fish. Next the state blamed the beaver. The Conservation Department said the beaver were warming up the water by damming tributary streams, and the department began dynamiting dam after dam, dam after dam, hundreds of them. Acid rain was never mentioned. We never even heard about acid rain until five years ago, when we started reading about the trouble in Sweden and Canada. Who'd ever think it would happen here?"

By his own account, Bowes should have known better. Before he bought the lodge, a rustic hotel, in 1951, he was a professional naturalist on the staff of the National Audubon Society. Even after Bowes left Audubon, he led field trips to Central and South America, the Caribbean and Africa, but for all his expertise in the natural world, he admits he didn't have a glimmer about acid precipitation until it was too late and he had experienced the effects firsthand. "I should have known," Bowes said, "but without excusing myself, most Americans don't know what is happening, and something new is happening every day with acid rain."

The incident that opened Bowes's eyes occurred in 1980, at the start of the summer tourist season. Every July and August, when business is hectic at Covewood Lodge, Bowes, his wife, Diane, and their two young daughters, Kimberly and Rebecca, move out of their house and live in an apartment on the third floor of the hotel. They are downstairs busy with guests all day long and use the apartment only at night. One evening in July, Kimberly and Rebecca turned on the faucet in the apartment to get a drink of water. They complained that the water tasted "funny." Indeed it did, and analysis disclosed that it contained five times the State Health Department's permissible amount of lead and three times the permissible amount of copper. Acids entering the spring that supplied the water for the hotel had leached the lead and copper from the plumbing in poisonous amounts. The lead and copper could be tasted in the water from the apartment faucet because it had been standing in the pipe all day. Bowes checked all the water pipes on his property. All but one had highly acid water. The exception was the pipe in his own house. He was puzzled by this until he learned that the contractor who had built the well serving the house had used limestone tiling, and the limestone was neutralizing the acid in the water. By constructing limestone filter beds for the Covewood Lodge water supply, Bowes was able to correct the problem, but the question remains: How many people dependent on water in the

Adirondacks, or other parts of the United States where acid precipitation comes into contact with metals, know what's in their water? And what is the consumption of that water doing to those who drink it?

Five miles up the road from Covewood Lodge in the hamlet of Big Moose, Bill Marleau sits in an armchair in the living room of his small frame house. Marleau hasn't been feeling well of late, but instead of talking about his own ailments he's ticking off the names of the lakes in the area that have been acidified, including Woods Lake where he has had a small cabin for years. Except for the three years he served in the Navy in World War II, Marleau has spent his entire life in Big Moose, and he is of the opinion that acid precipitation has had an even more widespread effect on the region than has been documented so far. Marleau is not a scientist. He doesn't have data printouts to substantiate his views. But he is a state forest ranger—he has been one for thirty-three years—and until the fishing collapsed, he was an ardent angler. What Marleau has to say is based on a lifetime of working, fishing, exploring, camping, hunting, and trapping in the region around Big Moose.

"Almost everything is down," said Marleau. "*Everything*. Acid rain affects the birds that feed on fish, the fur-bearing animals that feed on fish. The way I look at it, everything in nature is dependent on food, and when you reduce the food supply of those birds and animals, it affects other birds and animals that aren't directly dependent on aquatic insects and fish." According to Marleau, every pond and lake used to have a colony of forty to fifty tree swallows. Nowadays he is lucky to see a colony of six.

Often he sees no swallows at all. In late August when the swallows gather to migrate, he used to see a thousand of them on the phone wires near the old railroad station in McKeever. The last time he counted there were only eighteen. "The snowshoe rabbit is down, the fox is way down, deer are down, way down, the bobcat is down, the raccoon is down," Marleau said. "Even the porcupine is disappearing. Bear is fairly plentiful, but of course a bear is like a pig. It will eat anything from bark to garbage. Salamanders are down, both the red and the black with yellow spots. Frogs and crayfish are way down. Kingfisher, osprey, gulls—they're all down. The loon has disappeared. Grackles and blackbirds are down. Blue jays, tipups, the hermit thrush, white-throated sparrows, song sparrows are down. There are no mayflies, and darning needles are way down. There used to be clouds of them when there were thick hatches of mosquitoes. The mosquitoes are nothing like when I was a kid. I remember when those electric power fellers came from Ohio. They thought Woods Lake was beautiful, and I said, 'Yes, but you never saw it like I did. There were osprey and golden eagles here. I could see fish jumping. You don't miss the things you never saw before.'

"Now you don't see a fish jump anymore," Marleau continued. "There's no fish to jump and no insects to make them jump. It isn't a matter of trying to lime a lake and stocking fish. The fish have to eat, and they'd just eat one another because there's nothing else. It gets to a certain point where you're going to have to play God again and start all over with the littlest insect."

2

Slow-Motion Destruction

Although acid precipitation has been recognized as a problem in recent years, evidence indicates that it probably began on a very small scale with the advent of the industrial age. In an effort to reconstruct the history of acidification, Drs. Ronald B. Davis and Stephen A. Norton and colleagues at the University of Maine studied sediment cores taken from the deepest parts of twenty lakes in New England, thirty-eight in Norway, and ten in northern Sweden. Located in granitic areas with shallow soils, these lakes are particularly sensitive to atmospheric influences. Examination of cores extracted from all of them showed that lead and zinc, heavy metals associated with the combustion process, "started increasing detectably above preindustrial background concentrations several decades to more than a century ago." Zinc increased by as much as 700 percent over the background level, while lead increased 300 percent. Their studies of the sedimentary remains of diatoms, microscopic one-celled algae, and of cladocerans, microscopic crustacea, revealed that these organisms "demonstrated biological changes related to acidification in some of the lakes."

Acid rain as such was first described as a local phenomenon in England in 1852 by Robert Angus Smith who did his research in and around the industrialized city of Manchester. Smith blamed the sulfuric acid in the city air for the rusting of metals and the fading of colors in dyed goods. In 1872 Smith published a book, *Air and Rain*, in which he observed that coal burning in Great Britain caused acid rain. Shortly after the turn of the century, English scientists C. Crowther and A. G. Ruston reported that acid rain, the term they used, had killed or reduced the yields of timothy, radish, lettuce, and cabbage grown near Leeds.

As early as 1895, evidence now shows, acidification of lakes in southern Norway caused a decrease in fish populations, and by 1905 trout had become extinct in several Norwegian lakes, although no one knew the reason why. In 1948 Hans Egner, a Swedish soil scientist who was interested in the fertilization of crops by atmospheric-borne nutrients, set up a network of sampling buckets at experimental farms throughout the country to measure monthly deposits of rain, snow, and dust. Scientists elsewhere in Europe followed suit, and in 1956 the International Meteorological Institute in Stockholm began to coordinate the data being accumulated by the new European Air Chemistry Network. Two other Swedes, Erik Ericksson and Karl Gustav Rossby, who developed the science of atmospheric chemistry, were convinced that the atmosphere was responsible for the long-distance transport of many substances, and they used the data to test their ideas about the movements of air masses and the processes of atmospheric scavenging and deposition.

Also during the 1950s, Dr. Eville Gorham, a young Canadian ecologist then in England and now at the University of Minnesota, began publishing a series of papers that demonstrated that acid precipitation could affect the buffering capacity of bedrock, soils, and lakes. Indeed, Gorham and a colleague, F. J. H. Mackereth, found acidified high-mountain tarns in the English Lake District, and they deduced that the tarns were acid because of air pollution. Gorham observed that acid rain fell in the region whenever the wind blew from urban/industrial areas. Gorham also found acidified waters in Cheshire. Gorham also linked the incidence of bronchitis with acid rain, lung cancer with tar in the air of British cities, and the pneumonia death rate with suspended sulfates. In other words, Gorham associated three different diseases with three different components of air pollution. As Dr. Ellis Cowling, associate dean of North Carolina State's School of Forest Resources, wrote in a review of research on acid precipitation, "Thus, by the mid 1950s and early 1960s, Gorham had established a major part of our present understanding of the sources of the limnological and ecological significance of acidity in precipitation. But these pioneering researches were met by a thundering silence from both the scientific community and the public at large. This lack of recognition may have resulted in part because Gorham published his highly interdisciplinary researches in a diverse array of respected scientific journals. Whatever the reason, this lack of recognition resulted in at least a ten-year lag in both scientific and public awareness of acid precipitation and its ecological significance."

A thundering silence also greeted an article published in 1959 by A. Dannevig, a Norwegian fisheries inspector. In an obscure journal, *Jeger og fisker,* he advanced the idea that the acidity of precipitation was causing waters to become too acid for fish.

It was not until the late 1960s that Svante Odén, a young colleague of Rossby and Ericksson, made the breakthrough that identified acid precipitation as a serious environmental threat. Odén, who had been asked to do research on surface-water chemistry, theorized that the increasing acidity of Swedish waters was the result of atmospheric fallout of sulfates that had traveled long distances. The Swedish government asked him to write a report on his hypothesis, and while Odén was working on it, he received a call from a fisheries inspector in western Sweden who asked, "Is it possible that a massive fish kill we have found could be related to the acid precipitation?" Odén recalls, "That was a shock to me because that was the first real indication that acid precipitation had an impact on the biosystem."

Odén first published his idea in a Stockholm newspaper, *Dagens Nhyeter,* in 1967, which annoyed some scientists, and in 1968 he published an article, "The Acidification of Air and Precipitation, and Its Consequences in the Natural Environment" in the *Ecology Committee Bulletin*, which was translated into English that same year in the United States. "His analyses of air-mass trajectories clearly showed that acid precipitation was a large-scale regional phenomenon in much of Europe," Cowling wrote, "that both precipitation and surface waters were becoming more acidic, and that long-distance (100–2000 km) transport of both sulfur- and nitrogen-containing air pollutants was occurring among the various nations of Europe. Odén also enunciated a series of hypotheses about the probable ecological consequences of acid precipitation—decline of fish populations, decreased forest growth, increased plant diseases, and accelerated damage to materials. These conclusions and hypotheses led to a veritable storm of scientific and public concern about acid precipitation."

In 1969 Odén attended the 19th International Limnological Congress in Winnipeg, and in 1971 he lectured at a number of institutions in the United States. Scientists in this country, notably Dr. Gene E. Likens at Cornell and Charles Cogbill, one of his graduate students; Dr. F. Herbert Bormann of Yale; Dr. James N. Galloway of the University of Virginia; and Ellis Cowling at North Carolina State, began investigating acid precipitation. Cowling soon helped establish the National Atmospheric Deposition Program, which now maintains a network of sampling stations throughout the United States. At Cornell Dr. Dwight Webster, a professor of fisheries science who had been working on several Adirondack lakes that had lost their fish, asked Dr. Carl Schofield, an aquatic scientist, to examine the data Webster had accumulated, and as Webster recalled, "Everything began to fall into place." In Canada Drs. Richard J. Beamish and Harold H. Harvey of the University of Toronto published a paper, "Acidification of the La Cloche Mountain Lakes, Ontario, and Resulting Fish Mortalities," in 1972. That same year, the Norwegian government initiated the so-called SNSF Project, Acid Precipitation—Effects on Forest and Fish, which sought as its first objective to establish as precisely as possible the effects of acid precipitation on forests and freshwater fish. In 1980 the final report of the project, which involved scientists from other countries, concluded, "The recent acidification of freshwaters in parts of Europe and eastern North America has had profound impacts on aquatic life. It can be stated with reliability that all trophic levels are affected. Of immediate concern to the people living in the acidified regions is the major decline in fish populations. In the four southernmost counties in Norway more than half the fish populations have been lost during the 1940–1980 period. Today, lakes in more than 13,000 km^2 of south Norway are practically devoid of fish, and the fish stocks are reduced in an additional 20,000 km^2. Continued water acidification is a threat to hundreds of lakes still harbouring valuable fish populations. High egg and fry mortality in acid water is regarded as the main reason for fish decline, but other population responses are also known. Massive fish kill of adult fish during acid episodes is well documented, and is caused by physiological stress from toxic combinations of water acidity and aqueous aluminum."

As a result of research efforts in Norway, Sweden, the United States, and Canada, we now know a considerable amount about what increasing levels of acidity do to various species of fish. At pH 5.5, brook, brown, and rainbow trout experience significant reductions in egg hatchability and growth. At pH 5.5, largemouth bass, smallmouth bass, walleyes, and rainbow trout are eliminated, and declines in other trout and salmon populations can be expected. Below pH 4.8, lakes are generally fishless. Research in Norway indicated that of nine species observed there rainbow trout are the most sensitive to acidification, followed in order by Atlantic salmon, sea trout, brown trout, perch, char, brook trout, pike, and eel.

A low pH can cause female fish to retain their eggs, but even if the eggs are laid, mortality can be high in acidified water because fish are ultrasensitive in the egg, larval, and fry stages. At pH 4, the egg capsule of an Atlantic salmon will split, but the head of the larval salmon will remain stuck inside because of acid interference with chorionase, the hatching enzyme. Ironically in some lakes as new year-classes of fish fail to develop, the older fish become larger because of reduced competition for food, and anglers will report sensational catches just before acidification reaches lethal levels.

Why do the fish die off? Low pH by itself interferes with the salt balance that freshwater species need to maintain in their body tissues and blood plasma. But apart from that, there is another factor at work: aluminum mobilized by acid precipitation. Carl Schofield discovered that aluminum can be lethal to fish and other organisms at pH levels that are normally considered safe for them. Acidification also mobilizes manganese, zinc, nickel, lead, mercury, and cadmium. The concentration, or "bioaccumulation," of mercury in fish poses very serious problems. Those that do not die may become toxic to predators, including humans, who eat them. The mobilization and/or atmospheric deposition of metals opens a Pandora's box of poisons. In Sweden the government warned the public not to eat the livers and kidneys of moose. These organs are contaminated with cadmium ingested by the animals browsing on plants in acidified waters. Some livers are sufficiently poisonous to be fatal to humans.

In addition to loss of its fish life, an acidified body of water also loses hundreds of other organisms, including certain types of algae, crustacea, molluscs, and insects. For example, in a stream stoneflies and mayflies generally disappear at pH 5. Many species of these two important insect orders are detritivores that feed not on the submerged vegetation but on the dead leaves that have fallen or been blown or washed into the water. The leaf fall in autumn is one of the biggest transfers of energy on the planet, and these insect detritivores are a vital link in the energy flow that goes from the sun, soil, and trees to stoneflies and mayflies to trout to fish-eating birds and mammals, including man. And here we would like to make a point that has not been raised thus far by researchers: Eradication of rare aquatic insects can destroy part of the evolutionary record. For example, in the late 1950s,

Drs. Herbert H. Ross and William E. Ricker, two of the most distinguished aquatic entomologists in the world, looked around for an insect whose development and distribution could yield reliable information on the effect of the Pleistocene glaciers on the biota of eastern North America. They settled on stoneflies of the genus *Allocapnia,* and with the help of more than 150 scientist-collectors, they began their work. Unlike most insects, adult stoneflies of the genus *Allocapnia* emerge during the colder months of the year, and they can readily be seen trotting about on the snow next to a stream. In 1971 Ross and Ricker published *The Classification, Evolution and Dispersal of the Winter Stonefly Genus* Allocapnia, in which they reported that the genus had evolved about three or four million years ago, primarily in the Appalachian Mountains and neighboring ridges. Six ancestral lines spread west into the Ozark-Ouachita Mountains where they evolved into distinct lineages, one of which spread back eastward into the Appalachians. Ross and Ricker wrote, "The evidence at hand suggests that all the phylogenetic developments of the genus that we can deduce started late in the Pliocene when tectonic uplifts in central North America and subsequent intensified erosion had produced an avenue of spring-fed streams that allowed dispersal of the genus between the Appalachian and Ozark-Ouachita systems. The evidence further suggests that the speciation pattern of *Allocapnia* is associated with the alternation of cold glacial and warm interglacial periods of the Pleistocene and comparable climatic oscillations occurring in late Pliocene." This is valuable information to students of evolution, entomologists, climatologists, geologists, and others who may use it in ways that we cannot even guess, but the Appalachian streams where a number of these stoneflies were found during the study are under assault from acid precipitation. This particularly applies to three rare species—*A. brooksi, A. fumosa,* and *A. stannardi*—known only from a few locations in the Great Smoky Mountains, an area that is being especially hard hit. There is also a moral issue involved in the extinction of a species. As William Beebe wrote, "The beauty and genius of a work of art may be reconceived though its first material expression be destroyed; a vanished harmony may yet again inspire the composer; but when the last individual of a race of living beings breathes no more, another heaven and another earth must pass before such a one can be again."

A few species can live, even flourish, in an acidified body of water. A species in point is the water boatman, which brings to mind the time that Fred Johnson, Water Resources Coordinator for the Pennsylvania Fish Commission, visited an acidified lake in the Adirondacks with a busload of fellow Pennsylvanians, including apologists for the coal industry, one of whom flooded the bus with copies of a pamphlet called "Misconceptions About Acid Rain." Upon reaching the lake, one of the apologists leaped forward to scoop some water, examined the sample with a hand lens, and exclaimed, "Why there's life here! This lake is alive!" As Johnson remarked "That description of alive won't put fish in the creel."

The water in an acidified lake is often a beautiful crystalline blue, but the bottom is sometimes carpeted with fibrous mats of algae, thick enough to be picked up and shaken like a rug. Bacteria that can live without any oxygen thrive beneath such mats, where they decompose plant matter and produce gases that bubble up to the surface during the summer months. "I suspect that this is the cause of the garbage-dump-like odor that wafts over the surface of some acidified Adirondack lakes during the warmest part of the year," said George Hendrey of Brookhaven.

There is yet another bizarre touch: Tree leaves that fall into acidified ponds, lakes, and streams become pickled and simply stay there. The bacteria and fungi that normally would begin to break down the leaves are inhibited, and the stoneflies and mayflies that would eat leaf detritus are long gone. Given this, leaves can build up and choke a body of water, and as Hendrey has said, "Acidification is accelerating the filling-in of ponds. The accumulation of material is abnormal, and it's increasing so rapidly that it may soon have negative effects for human beings."

In Canada Dr. David Schindler heads a team of scientists at the Freshwater Institute in Winnipeg who are attempting to document the most minute changes that occur in acidified lakes. The Canadian government in 1969 established a unique area, the Experimental Lakes Area, southeast of Kenora, Ontario, by setting aside forty-six lakes for scientific investigation of pollution. Like thousands of other lakes in eastern Canada, the ELA lakes are situated on the granitic Canadian Shield, but because they are remote from sources of pollution they are basically undisturbed. Inasmuch as the hydrological, meteorological, chemical, biological, and physical characteristics and parameters of all forty-six lakes have been measured and noted, one of them can serve as an experimental laboratory, with others acting as controls. Early research in the ELA centered on the effects of phosphate detergents, which led to their reduced use in Canada. In 1976 the emphasis shifted to the effects of acid precipitation. Schindler and his colleagues took one lake—it has no name, simply the designation Lake 223—and began adding sulfuric acid to it to lower the pH from the original level of 6.8. Here, according to Dr. Ken Mills, a fisheries biologist, is what has happened over the years.

In 1977 the pH was 6.1, and there was an increased emergence of diptera, flies that serve as a major food for white suckers. In 1978 the pH was 5.8, and a species of copepod, a small crustacean that serves as food for fish, disappeared. Fathead minnows failed to reproduce, and there was a possible increase in the embryonic mortality of lake trout. In 1979 the pH sank to 5.6, and dense algal mats developed along the shoreline. A mysid shrimp, *Mysis relicta*, an important food for lake trout, disappeared; the slimy sculpin declined; the population of fathead minnows declined severely; and crayfish, *Orconectes virilis*, had softer shells as the calcification of their exoskeletons decreased. In Scandinavia, a once-common crayfish, *Astacus astacus*, has become rare in lakes with a pH below 6.

In 1980 the pH of Lake 223 was 5.4, and another species of copepod disappeared. A species of *Daphnia*, ordinarily eaten by mysid shrimp, increased in numbers. There was an increased infestation of crayfish by a microsporozoan similar to those parasites that cause malaria, and the muscle color of crayfish tails turned white. Three-quarters of the adult crayfish disappeared, and those that remained suffered impairment of reproduction. There was a greater incidence of fewer eggs attached to female crayfish. Ordinarily, after a female crayfish extrudes the fertilized eggs, she cements them in bunches like berries to the swimmerets on her underside until they hatch. It is believed that the increasing acidity of the lake changed the mechanism attachment. Reproductive failure also affected lake trout, but there was an increased abundance of the pearl dace, a minnow, that filled the niche left vacant by the disappearance of the fathead minnow.

In 1981 the pH was 5.1, and there was a recruitment failure of white suckers. The fish laid eggs, but young white suckers did not appear. Any fisherman who chanced to happen upon Lake 223 would still find adult lake trout and white suckers and think that things seemed just fine, but the demise of these two species has been set in motion by reproductive failure.

This slow-motion destruction is happening to other lakes and streams where the acid is not coming from scientists but from the polluted sky. In March 1982, Dr. Orie L. Loucks of the Institute of Ecology in Indianapolis released a study that the institute had conducted for the Congressional Office of Technology Assessment. The study estimated that approximately 9,400 lakes, plus or minus 20 percent, in the eastern United States had alkalinity levels of less than 200 microequivalents per liter, "a level indicating either sensitivity to acid inputs or existing alterations by these inputs." In addition, 51,000 miles of streams in the eastern United States "can be categorized as aquatic resources already altered or seriously at risk."

3

Damaged Waters

From high in a jet flying over eastern Canada, the view is one of endless forests laced with ponds, lakes, streams, and rivers. This is the dream world of the North Woods come true. Three-quarters of northern Ontario and nine-tenths of Quebec, the largest of all the Canadian provinces, lie across a glacially scoured geologic formation called the Canadian Shield, a granitic sheet of ancient Precambrian rock. The poor soils of the shield offered the first settlers a harsh life of toil, but livelihoods—even fortunes—were later to be found in logging, trapping, and fishing. Waters teemed with smallmouth bass, trout, and Atlantic salmon. This was the land of the Voyageurs, a country of pristine streams and clean, cool, bracing air. Here in the La Cloche Mountains, north of Sudbury,

Ontario, the artists known as the Group of Seven painted the unspoiled hills and lakes that became the signature of Canada. Now this dreamworld is turning into a nightmare, in part because of Canadian abuse of the environment, but also in good part because of acid-laden emissions coming from the United States. If a hostile power were doing to us what we are doing to Canada, the Congress would now be meeting to consider a declaration of war.

In Canada acid precipitation has now been reported as occurring from Alberta eastward to Newfoundland. With the exception of parts of New Brunswick and all Prince Edward Island, most of the rest of eastern Canada is considered susceptible to acid precipitation.

Ontario. An estimated 1,200 lakes have lost populations of fish because of acidification, approximately 3,400 more are close to that state, and 11,400 are considered at risk. That is just the beginning, according to the Ontario Ministry of the Environment. At the current rates of deposition, an estimated 48,500 lakes in the province are likely to lose their fish in the next twenty years. Among the lakes threatened are those in the Muskoka-Haliburton region, the heart of Ontario's tourist industry. The mean pH of rainfall in the region now ranges from 3.9 to 4.4.

Quebec. In a brief submitted to the U.S. EPA in 1981, the Quebec Ministry of the Environment called for reductions of sulfur dioxide emissions coming from sources in Ohio, West Virginia, Illinois, Indiana, Michigan, and Tennessee because of the harm they were doing to the province, especially that part situated on the granitic Canadian Shield north of the St. Lawrence River. About 60 percent of the pollution from sulfur compounds that affect the province originate in the United States, the brief said. Testing of lakes in Quebec has thus far revealed that 17 percent were acidified, 39 percent extremely sensitive, and 28 percent moderately sensitive. The brief argued that large quantities of ions in lake waters were clear evidence that Quebec was receiving acid fallout because the bedrock in the province is very low in sulfates. In capital letters the brief noted: "WE HAVE EVERY REASON TO BELIEVE THAT ACID PRECIPITATION HAS CAUSED WIDESPREAD INCREASES IN ACID LOADINGS OF LAKES IN QUEBEC'S PRECAMBRIAN SHIELD AREA." Indeed, acid loadings in lakes in watersheds feeding the Outaouais and St. Lawrence Rivers and the Gulf of St. Lawrence were on the same scale as those for lakes acidified in south-central Ontario and southern Norway. Furthermore, the Ministry of the Environment said it had statistical evidence in hand that demonstrated that watersheds with high acid loadings contained the greatest number of lakes with abnormally low catches of fish. Indeed between 1970 and 1978, the annual catch in more than 150 lakes in the Parc des Laurentides north of Quebec City, an area particularly vulnerable to acid fallout, dropped 30 percent.

The brief concluded, ". . . Quebec is faced with a Gordian knot in the form of pollution of air masses originating mainly south of our borders, and . . . we are compelled to submit to gradual acidification of our province by a constant flood of atmospheric pollutants. Recent studies indicate, to our dismay, that acidification of Quebec's lakes and its impact on aquatic life is highly advanced, and this, in spite of the fact that we have only just begun to scratch the surface of the problem. We in Quebec still know too little of the effects of air pollution on our vegetation, agriculture, and forest resources, but we are aware of the disastrous effects it can have through our knowledge of the studies and data collection carried out in regions and countries with geology and climate similar to our own."

One of the big questions in Quebec is, what is acid precipitation doing to the sixty Atlantic salmon rivers on the north shore of the St. Lawrence? In 1981 Canada's Department of Fisheries and Oceans began studying four of them—the Petit Saugenay; the St. Marguerite, which is a tributary of the Saugenay; the Escoumins and the Petit Escoumins. The data gathered thus far show no apparent effect from acid precipitation, except for the St. Marguerite. The river has high levels of aluminum, with a mean of 165 micrograms per liter. Although the pH varies from 6.4 to 6.7 in the St. Marguerite, the level of carbonates or the total alkalinity ranges from a low of 0.4 milligrams per liter in May after the snowmelt to a high of 4.0 in August.

There are some ominous signs in the watershed of another north shore river, the Moise, which is generally regarded as the greatest Atlantic salmon river in North America, at least in terms of size of the fish. "The average fish weighs more than twenty pounds," said Dr. Karl Schiefer, an Atlantic salmon expert and president of Beak

Consultants Ltd., an environmental firm in Toronto. The best fishing is controlled by the exclusive Moise Salmon Club, and an invitation to fish the club water is treasured. Guests have included the likes of Dwight Eisenhower and Viscount Alexander of Tunis. Dr. Schiefer, who has been studying the Moise for the last twelve years, said, "The pH and the total alkalinity of the main stem are at their normal background levels, but some tributaries are showing pH sags." Some of the smaller tributaries, such as the Taoti River, have experienced a pH drop below 5. Dr. Schiefer added, "I must qualify what I say now because it's based on one or two spot samples, but I've seen some pH problems in the Nipissis River and the Ouapetec River, two major tributaries of the Moise that produce a lot of salmon."

There is yet another potential threat to the Moise that could be exacerbated by acid precipitation. Iron mining has begun at Mount Wright, and the tailings are being dumped into a headwater tributary of the river. These tailings have high levels of mercury, and acid precipitation can make the mercury biologically available. Dr. Schiefer warned, "We are seeing elevated levels of mercury in the sediments. There it can enter the food chain. The major part of the diet for juvenile trout and salmon is benthic invertebrates—mayflies, caddisflies, and stoneflies—and when fish eat them they can wind up doubling or tripling the contamination levels that exist in the invertebrates." In the summer of 1982, Dr. Schiefer planned to do mercury analyses of fish to see whether, and to what extent, the contamination may have spread.

Nova Scotia. According to Dr. Walton D. Watt of Canada's Department of Fisheries and Oceans in Halifax, ten Atlantic salmon rivers in the province have become acidified and are unable to support fish. Eight of these rivers—the Tusket, Barrington, Clyde, Roseway, Jordan, Sable, Broad, and Larrys—have a pH below 4.7. Two other salmon rivers that have lost their fish, the Nine Mile and the Salmon—the latter the one near Lawrence Town (Nova Scotia has several rivers with the same name), have a pH below 5. According to Watt, "For several of these rivers there are angling records going back over a hundred years which show fairly steady catches until the 1950s, when a decline set in, and in most cases no angling catch by 1970. An electrofishing survey in the summer of 1980 failed to find any sign of Atlantic salmon reproduction in . . . [nine rivers then sampled]; hence their salmon runs are now considered extinct."

In another eleven rivers with the pH ranging from 5 down to 4.7, Atlantic salmon still maintain self-sustaining runs, but their numbers have declined. These rivers are the Sissiboo, the Mersey, the East (near Chester), the Middle (also near Chester), Ingram, Sackville, Tangier, the West (near Sheet Harbour), Liscomb, the Isaacs Harbour River, and the New Harbour. In addition, there are nine more rivers with the pH between 5.1 to 5.4 where Atlantic salmon are considered "threatened." These rivers are the Gold and the Medway, both especially significant for anglers, and the Salmon (near Yarmouth), the East (near Sheet Harbour), Port Duffern, Moser, Bear, Gaspereaux, and the Cole Harbour. Assuming that the rate of acidification continues unchanged, Watt predicts the extinction of the salmon runs in eleven more Nova Scotia rivers and run depletion in another nine by the end of the century. "It seems unlikely that the rate of acidification will continue unchanged," Watt said. ". . . all indications are that it will accelerate."

Newfoundland and Labrador. The pH of thirteen streams being monitored in eastern Newfoundland varies from 5.3 to 6.1, "close to levels at which fish populations could be threatened," said David G. Jeans, assistant deputy minister of the environment. Most inland waters in the province are sensitive to acid precipitation. In Newfoundland the mean pH of precipitation ranges from 4.5 at Stephenville to 4.9 at Gander. In Labrador the precipitation pH increases from 4.7 in the south to greater than 5 in the north. Jeans thinks it would seem unlikely, "although not impossible," that there are already acidified lakes in the province now, but "further increases in acidic pollutants being transported into the Province will seriously alter this situation, thereby causing a decrease in salmon and trout productivity in many water bodies in Newfoundland and southern Labrador."

New Brunswick. Research on the effects of acid precipitation has yet to get under way, but the province has sensitive areas and Premier Richard Hatfield is very much concerned. "I'm not worried about today," he said. "I'm worried about ten years from now. Our forests take between thirty and thirty-five years to mature, and I think the evidence is that acid rain is going to stunt the growth of trees. The federal government in Ottawa has to become more aggressive on the provincial level, and Washington has a political-ecological problem it must solve. They [the Reagan Administration] believe in giving industry free rein, and that doesn't work in situations like this."

The picture of acidified waters is just as bleak in various parts of the United States, especially in the Northeast, which, like eastern Canada, serves as the atmospheric garbage dump for the continent.

Maine. Dr. Terry A. Haines of the U.S. Fish and Wildlife Service in Orono reports that Atlantic salmon rivers east of the Penobscot River have lower pHs and lower total alkalinities than those west of the Penobscot. The Union River has a pH of 6.6; the Narraguagus, 6.5; the Pleasant, 6.4; the East Machias, 6.2; and the Denys, 6.1. A small tributary of the Machias, West Kerwin Brook, had a pH of only 5.3. "As long as salmon fisheries are maintained by the stocking of hatchery-reared juveniles or smolts, United States salmon populations are unlikely to be affected by acidic precipitation," Haines said. "However, some small tributaries are apparently already sufficiently acid to affect survival of salmon fry. If present trends continue, the reestablishment or continued existence of self-sustaining populations could be jeopardized."

New Hampshire. The state has yet to do a complete survey of its high-altitude lakes, which are likely to suffer first, but Ronald Towne of the Water Supply and Pollution Control Commission, said, "The usual picture of acid-pickled lakes is beginning to emerge. We have lakes with low pH, low alkalinities, high levels of aluminum, no fish or missing year-classes of fish, no salamanders, no frogs. We're as vulnerable as any state, and we are really getting hurt." Thus far, fifteen high-altitude lakes that Townes was able to reach by car are "bad," but he has been unable to sample remote waters because the commission lacks the money for a helicopter.

Vermont. Several lakes in the Brooks Wilderness Area of the Green Mountain National Forest have a pH of 4, and two tributaries of the West River, Ball Mountain Brook and the Windhill River, have acid headwaters. A 1979–1980 survey of 120 lakes in the state revealed that the majority of them were potentially susceptible to acidification because their total alkalinities were less than 20 ppm. In 1979 the town of Bennington had to spend $37,000 for large amounts of sodium bicarbonate and sodium hydroxide to reduce the corrosivity of the drinking water drawn from Bolles Brook, a small mountain stream. The raw water was often below the federal drinking-water standard of pH 6, and some houses in Bennington with lead water pipes had high levels of lead in their tap water. (Acid precipitation and associated heavy metals may be responsible for a 50 percent dieback of red spruce in the Green Mountains near Burlington. There have also been dramatic declines in beech and sugar maples. These declines are discussed in the next chapter.)

Massachusetts. Fisheries and drinking-water supplies are both facing potential disaster. In late July of 1981, the pH of a rainstorm in Lawrence, Massachusetts, was 2.9. More than 60 percent of 154 reservoirs surveyed in the state in 1980 are vulnerable to acid precipitation. The main concern is the Quabbin Reservoir, the source of $1 million-a-year sports fishery and the water supply for two million people in the metropolitan Boston area. According to Alan VanArsdale, the head of the Massachusetts Acid Deposition Assessment Program for the Department of Environmental Quality Engineering, the pH of the surface water often drops into the 5's and sometimes into the 4's. In the springtime, the pH of bottom water in the Quabbin goes below 5.5. Total alkalinity is very low, ranging from one to four parts per million in different parts of the reservoir as compared to eight to twelve parts per million in 1965. In the last six years, the levels of sulfur appear to be rising, at least in the surface waters. One of VanArsdale's main concerns is that acid drinking water can corrode pipes. In 1974 a sampling of Boston tap water revealed lead concentrations up to five times higher than the level permitted. The problem was solved by treating the water to raise the pH. In 1980 alone, the Metropolitan District Commission spent $469,000 on chemicals to raise water pH in the Boston area. Given the present trend of acid precipitation, Dick Keller of the state's Division of Fish and Wildlife fears that the Quabbin will lose its fish populations sometime in the next decade.

Other trouble spots: North Wattupa Pond, the reservoir for Fall River; Atkins Reservoir, which serves Amherst; and a series of reservoirs and ponds in the Dalton-Hinsdale area in the Berkshires. "There may not be fish in them anymore," said VanArsdale of the Dalton-Hinsdale ponds, which are located at elevations of 1,200 to 2,000 feet.

Plymouth County and Cape Cod in the southeastern part of the state have their problem waters. "Some ponds in Plymouth County are going downhill, some are not," Keller said. "It depends on the local geology. Some ponds have no natural reproduction of fish, but whatever the case they all face trouble since the total alkalinities are only between eight and fourteen parts per million." On the Cape, liming has long been a necessity because of the natural acidity of the soil, but the current levels of acid precipitation have apparently rendered Hathaway Pond in Barnstable beyond help, and it is no longer limed or stocked with trout. "There's literally nothing there now," said James Kennedy, district manager for the state's Division of Fish and Wildlife. In the winter of 1980–81, snow on the Cape had a pH of 3.4.

Rhode Island. The total alkalinity of the Scituate Reservoir, which serves as the drinking-water supply for half the state, is low, ranging from three to seven parts per million. In the summer of 1981, rainfall in Rhode Island averaged a pH of 3.5.

New York. The N.Y.S. Department of Environmental Conservation (DEC) has documented the acidification of 212 Adirondack lakes and ponds, totaling some 10,460 acres, that are incapable of supporting fish life. Most of these lakes and ponds are at an elevation of 2,000 feet or more in the southwestern Adirondacks, the first high ground to intercept polluted air masses coming from the Middle West. It is important to note that this figure of 212 is derived from tests made on only a third of the lakes and ponds in the Adirondacks. From this same limited sampling, another 256 lakes and ponds totaling some 63,000 acres were judged to be in danger of losing their fish. In November of 1981, the DEC reported that 40 percent of 114 Adirondack streams sampled during and after the 1980 snowmelt showed "critical" or "endangered" levels of acidity. Fifteen percent of the streams were deemed critical because they had a pH of less than 5. No fish were found in these streams. One quarter of the 114 streams were termed endangered because they had a pH of 6 or less. Fish were absent in a number of these endangered streams. It is significant that most of the streams showing the highest levels of acidity were located in the southwestern Adirondacks.

The Adirondack headwaters of the Hudson River have become acidified in part, and according to Dr. George Hendrey of the Brookhaven National Laboratory, the main stem of the Hudson from its source in Lake Tear of the Clouds to Saratoga, 140 miles downstream, has lost "significant" total alkalinity in the last twenty years. Other sensitive areas in the state include the Tug Hill Plateau to the west of the Adirondacks, the Catskill Mountains, the Shawangunk Mountains, the Hudson Highlands, and Long Island.

Connecticut. It has been determined that a dozen lakes have a total alkalinity of less than five parts per million, but Charles Fredette of the state's Department of Environmental Protection has called acid precipitation a "long-range" concern, adding, "We don't have high-altitude lakes like New York or New Hampshire, and we have relatively good buffering capacity." This view is questioned by some authorities in the field. Moreover,

Connecticut officials have adopted the view, first advanced by I. T. Rosenqvist in Norway in 1978, that acidification of poorly buffered waters is caused not by acid precipitation but by changing land-use practices. However, studies of Norwegian watersheds, some with changing land-use practices and others without, have shown that both are acidified at about equal rates. In addition, studies of North American lakes in areas where land-use practices have never changed have shown losses in buffering capacity. In short, Rosenqvist's theory is now discounted.

New Jersey. Research is just getting under way, but there are "some waters in the northwestern part of the state that show some signs of acidification," said Dr. Dean Arnold of the U.S. Fish and Wildlife Service. According to A. H. Johnson of the University of Pennsylvania, headwaters of streams in the Pine Barrens show signs of acidification from precipitation. In 1966 the U.S. Geological Survey reported that the 150-square-mile Pine Barrens "have no equal in the northeastern United States, not only for magnitude of water in storage and availability of recharge [from precipitation] but also for the ease and economy with which a large volume of water could be withdrawn." The survey reported the chemical purity of the water "approaches that of uncontaminated rainwater or melted glacier ice." (Johnson and Dr. Thomas Siccama of Yale University have also found "an abnormal decrease in growth rates for the past twenty-five years" for both pitch pine and loblolly pine in the Barrens. Their findings are discussed in the next chapter.)

Pennsylvania. "At present perhaps half our mountain streams can no longer support rainbow trout, and about a quarter of our first- and second-order streams can't support brown trout," said Fred Johnson, Water Resources Coordinator of the Pennsylvania Fish Commission. "There are also streams that we cannot stock before the trout season begins because of the acidity of the snowmelt. The situation is very serious." The central and northern parts of the state routinely have the most acidic rainfall of any large area in the country. The summer average is pH 3.8. It is a matter of irony to beer-drinking fishermen that Rolling Rock Creek in the Laurel Mountains near Latrobe, the stream from which Rolling Rock beer takes its name, is being acidified. TV commercials for the Latrobe Brewing Company feature the stream as a sylvan gem, but beer drinkers can take some small solace in learning that the brewery actually takes its water from another source.

Maryland. The average pH of precipitation in the western part of the state is 4, with individual storms having a pH as low as 3.4. Deep Creek Lake, the only major lake in the mountainous western part of the state, is considered "on the brink" of losing its fish.

West Virginia. A dozen trout streams are too acid to support fish. Moreover, 150 miles of the state's total of 550 miles of native brook trout streams are considered "threatened," said Don Gasper of the Department of Natural Resources. "The average pH of this 150 miles of streams is 5.5, and in the springtime it dips down to 4.8 or 5, and then climbs up to 6 in September. If the stream pH were to decline a half of a pH unit, there would be no more fish." Stocked streams are also being affected. Gasper said that about 150 miles of stocked streams are too acid in the spring to be stocked. "West Virginia is a stream state, and we're talking about losing one quarter of our heritage," Gasper said. "What's coming down is very, very bad."

Kentucky. In Cumberland State Park in the western part of the state, acid deposition is leaching heavy metals into watersheds.

North Carolina–Tennessee. The Great Smoky Mountains National Park, which covers 500,000 acres in both states, is taking a battering. The beautiful blue haze that comes from lacquers and oils liberated from the forest canopy is rarely seen. Instead visibility has been greatly reduced by an ugly gray haze composed of fine man-made particulates, mostly ammonium sulfate. After the Los Angeles Basin, the western slope of the southern Appalachians, from Georgia north to Kentucky, has the highest frequency of air stagnation in the United States. In grim recognition of this, the Blue Ridge Parkway is now unofficially called the Gray Ridge Parkway.

Over the years the annual pH of precipitation in the park has gone from 5.3 in 1955 to 4.4 in 1973 and 3.7 in 1980. Brook trout, the only native salmonid in the park, are threatened with extinction. During the spring, stream pH levels drop to as low as 4.3, and there are increasing concentrations of aluminum. In one stream, Beech Flats Creek, zinc has reached nearly toxic levels for fish, and rainbow trout in the park contained more than permissible amounts of mercury in fish allowed for human consumption—that is, until the Food and Drug Administration raised the level from .5 parts per million to 1 ppm. In April 1981 a sudden surge of acid streamwater coming from the park killed 1,000 rainbows held in raceways at a U.S. Fish and Wildlife Service hatchery on the Cherokee Indian Reservation in Cherokee, North Carolina. In lakes lying just outside the park boundary, smallmouth bass have abnormal backbones, a condition generally associated with aluminum toxicity.

Amphibians, particularly salamanders, are also threatened. The park contains the greatest diversity of salamanders in the world, including the Plethodontidae, the lungless salamanders, which probably evolved in the region. The index organism for the park, which is one of the world's key biological preserves, is the lungless shovel-nosed salamander. Like other lungless members of this unusual family, the shovel-nosed salamander must breathe through its skin and the lining of the mouth. In the increasingly acid Smokies, the shovel-nosed salamander is coming under great stress. Because it needs to keep its skin clean, it ordinarily molts it once a day and then eats it to gain energy. Now, however, laboratory experiments show that acid conditions can cause shovel-nosed salamanders to molt their skins six to eight times a day and not eat them, R. C. Mathews, Jr., a park service biologist, reported. Moreover, their eyes often turn opaque from acidity.

The Great Smoky Mountains National Park contains 1,300 vascular plants, more than are found in England and perhaps all of Europe. There are sixteen trees of world-record size, and the single largest red spruce forest in the world is in the park. The needles of white pines already show burn damage from ozone, which reacts synergistically with acid precipitation.

South Carolina. The National Wildlife Federation (see Appendix 1) has rated the state's freshwater fisheries as "extremely vulnerable to acid rain effects," but we are not aware of any research going on within the state.

Georgia. Northeastern Georgia, extending from Rabun County to Pickens County, has low buffering capacity, according to state environmental officials. There are reports of skeletal deformities in smallmouth bass in Lake Chatuge, a reservoir, and officials say there is some indication that the deformities might be caused by the effects of low pH levels.

Florida. Acid precipitation threatens numerous poorly buffered lakes in the sandy central highlands region that runs the length of the peninsula. According to Dr. P. L. Brezonik, a water resources specialist formerly at the University of Florida and now at the University of Minnesota, the acidity of Florida rainfall has increased markedly in the last twenty-five years, with the most acidic rains, averaging below pH 4.7, falling on the northern two-thirds of the state. The two most acidic lakes studied, McCloud (pH 4.7) and Anderson-Cue (pH 4.8), contain little aluminum and still have populations of fish, including Florida largemouth bass. The fish, however, are badly stunted and emaciated because of lack of food.

Michigan. Some 16,000 lakes, each more than ten acres in size, are considered susceptible to acid precipitation. More than half the 8,000 lakes and ponds in the Upper Peninsula have an alkalinity of only about ten parts per million. Scientists at Michigan Technological University report that the spring thaw is likely to have the greatest single impact on the Keeweenaw Peninsula on Lake Superior. The peninsula receives one of the heaviest snowfalls in the United States, averaging about twelve to thirteen feet in the course of a winter, and the median pH for snowfalls in the winter of 1977–78 was 4.5. Previously it was assumed that the Keeweenaw Peninsula received "clean" snow coming from the Northwest, but evidence gathered by the MTU scientists indicates that winter air trajectories often come from industrialized areas ringing the southern Great Lakes from Chicago eastward to Buffalo to Toronto. One snowfall, a "contaminated event," occurred on March 14, 1978, when air trajectories from Chicago, Detroit, and Sudbury, Ontario, all converged on the peninsula, or what the scientists called "the target area." The pH of the snow was 3.9.

Wisconsin. Twenty-six hundred lakes, each more than twenty acres in size, are considered very susceptible to acidification because they have a pH of 6 or less and little alkalinity.

Minnesota. The northern part of the state, particularly the Boundary Waters Canoe Area Wilderness, is susceptible to acidification. In fact, the "Transboundary Air Pollution Interim Report" prepared in February 1981 by a work group of U.S. and Canadian scientists noted that "Atmosphere loadings near the BWCAW are at levels associated with the onset of lake acidification in Scandinavian countries. Because of this, it is likely that the most vulnerable lakes are already being affected by acidity from atmospheric sources."

Dr. Gary Glass of the EPA's Environmental Research Laboratory in Duluth believes that the acidity of the precipitation in Minnesota varies in wet and dry years. In time of drought, wind-blown alkaline soil particles from the Dakotas can offset the acidity of precipitation in Minnesota, but these particles are suppressed when the Dakotas receive heavy snow and rain.

Colorado. Acid precipitation is falling in the high Rockies northwest of Denver. Drs. William M. Lewis, Jr., and Michael C. Grant of the University of Colorado discovered this accidentally in 1975 while they were working 9,000 feet up at the university's mountain research station at Como Creek, adjacent to the Indian Peaks Wilderness Area. In the four years from 1974 to 1978, the pH of precipitation dropped at a rapid rate, from 5.4 to 5.0 to 4.8 and to 4.7. According to Lewis and Grant, "High mountain lakes of the Rockies tend to be poorly buffered because of the predominantly granitic substratum and thus are likely to be comparatively sensitive to changes in precipitation pH. Aquatic systems, especially lakes, will probably be affected first and most adversely, much in the fashion of the Adirondacks. The difficulties presented by acid precipitation in the Rocky Mountain region may well be intensified by the anticipated development of fossil-fuel reserves, especially coal and oil shale, in this region. The potential for change is large."

Some investigators have attributed the acid precipitation found in the Como Creek area to polluted air moving upslope from the Denver area, but in August 1981 Dr.

John Harte at the Rocky Mountain Biological Laboratory reported that small lakes and streams in the Elk Mountains near Crested Butte in western Colorado have very high levels of acid. Harte said that the pH of rain and snow in the area sank as low as 3.6, and he added that preliminary calculations indicated that the state's acid-precipitation problem might be caused by emissions from coal-fired power plants in northwestern Colorado and northern New Mexico.

New Mexico. Acid precipitation with pH levels often in the 4's and occasionally in the 3's has been reported in the Tesuque Watersheds in the Santa Fe National Forest.

Wyoming. The average pH of precipitation falling on Yellowstone National Park in 1980 was 5.2.

Montana. The pH average for precipitation at Glacier National Park in 1980 was 4.9.

Idaho. The 1980–1981 average at Craters of the Moon National Monument was 4.8. Note: The Idaho, Montana, and Wyoming averages are for wet deposition only. Dry deposition may exacerbate the problem.

Washington. Twenty-four of the sixty-eight lakes sampled in the Cascades and the Olympic Mountains by Drs. Eugene B. Welch and William H. Chamberlain of the University of Washington had a pH of less than 6. Seven lakes, all located in the Alpine Lakes Wilderness in the Cascades due east of Seattle, had a pH of less than 5.5. In a report submitted January 1981 to the National Park Service, Welch and Chamberlain noted that rainfall monitored in Seattle ranged from pH 5.2 down to 4.2 approximately 70 percent of the time. Their report added,

"Although some western Washington acidity in precipitation no doubt comes from nitric acid through gasoline combustion in metropolitan areas, most probably originates from a coal-fired power plant and a nonferrous smelter located in southern Puget Sound lowlands." Welch and Chamberlain recommended that more lakes be sampled to determine the number affected and that acidified lakes be examined closely to determine the extent of damage.

California. Kathy Tonnessen and John Harte of the Berkeley Lawrence Laboratory have found high-altitude lakes in the Sierra Nevada with a pH of 5.8 and high levels of aluminum, but it is not known if the levels are natural or man-made. Dr. Doug Lawson of the state's Air Resources Board said, "So far we have not been able to document any effects of acid precipitation on California ecosystems. We do know that we have the potential for environmental insult. The state has levels of acid precipitation as high or higher than any place in the country, and we do have areas that are very susceptible in the Sierra Nevada and around Los Angeles where there are exposed granitic surfaces. Acid precipitation is severe, but we don't know what the results will be. Compared to the Northeast, the problem is new." The lowest pH recorded for any precipitation in the state occurred in Pasadena in September 1978. The pH of a light drizzle measured by Dr. James Morgan of Cal Tech was 2.9. When scientists recently flew a plane through smog over Los Angeles, they were unable to conduct tests because acids had corroded the instruments aboard.

4

Other Effects

Acid precipitation is not simply one evil genie let out of the stack; it is a host of them. And we are only starting to glimpse the potential for damage.

With a few exceptions, the impact of acid precipitation on crops and forests has yet to be scientifically determined under natural conditions. This is because there are more variables to contend with in terrestrial ecosystems than in aquatic systems. Even so, work with simulated acid rain in laboratory, greenhouse, and field experiments has shown that it can predispose plants to infection by bacterial and fungal pathogens and lessen their resistance to insects, accelerate erosion of waxes on leaf surfaces, induce lesions on foliage, and inhibit nitrogen fixation by symbiotic bacteria— among other effects. Simulated

lated acid rain has caused reduced growth and/or reduced yield for such crops as lettuce, soybeans, pinto beans, red kidney beans, tomatoes, cabbages, peppers, beets, carrots, radishes, broccoli, and mustard greens. Other crops found to be susceptible to injury are alfalfa, rye, and oats. To repeat, this work has been experimental, but as Ellis Cowling has noted, "Substantial damage to crops in certain regions of the United States is [already] caused by the
ition of toxic gases, aerosols, and particulate
lfur dioxide, ozone, oxides of nitrogen, fluoride,
gen chloride cause serious damage to crops that
onsidered together with the possible effects of
sition." Indeed, the Congressional Office of
y Assessment recently estimated that ozone
alone is causing losses of from $2 billion to $4.5
ear in production of soybeans, wheat, corn, and
n the United States.
the Brookhaven National Laboratory, Dr. Lance
ew plots of soybeans that were exposed to
acid rain with different levels of acidity. Rain
4 caused a 2.6 percent reduction in seed yield,
ggested that current acid levels in rainfall over
heast have the potential of causing a $129 million-
ss to farmers, based on the 1979 market price. "The results of this experiment are to be considered preliminary, and another experiment is under way to confirm the findings," George Hendrey of Brookhaven testified before the N.Y.S. Department of Environmental Conservation in 1981. "We do not know if actual crop losses are occurring. It would be extremely difficult to observe a real

crop loss of two percent, which could be attributed to acid precipitation, because the natural year-to-year variability of crop yields are larger than this. . . . Nevertheless, such a loss would be real and would contribute to an overall reduction in the annual harvest and to substantial economic losses."

A 1980 study on emissions from coal-fired power plants in the Ohio River Basin estimated that acid precipitation was causing the leaching of essential nutrients and minerals, which resulted in a reduction of forest growth in Ohio Basin states by 5 percent a year. It is sometimes said that acid precipitation can have beneficial effects on forest soils on a short-term basis, but studies indicate that over the long term the effects can be damaging. In the Adirondacks the leaching of soil nutrients exceeds any contribution from precipitation.

Probably the most significant work on North American forests is that being done in the Green Mountains of Vermont by Dr. Hubert Vogelmann, Dr. Richard Klein, and Margaret Bliss of the Botany Department at the University of Vermont and by Dr. Thomas E. Siccama of Yale University's School of Forestry and Environmental Studies. In a paper scheduled for publication in the *Bulletin of the Torrey Botanical Club*, Vogelmann, Bliss, and Siccama report a startling 50 percent dieback in red spruce, *Picea rubens*, on Camels Hump, Jay Peak, Bolton Mountain, and Mount Abraham. The dieback has occurred in the last fifteen years on land that the scientists had previously studied and from where they had taken soil samples. "Examination of dying trees has not revealed disease organisms," they wrote. "The fact that trees of all ages become necrobiotic suggests that they are under environmental stress, but it is not clear what stress or stresses are involved. . . . Red spruce decline is especially pronounced at upper mountain elevations where precipitation is high and fog is of frequent occurrence. Studies currently under way in the Green Mountains indicate that both rain and fog at these elevations are highly acid. . . . Heavy metals (i.e., lead, copper, and zinc) are known to be accumulating in forest soils at upper elevations. Since the environment of high elevations is normally fragile, it is possible that recent atmospheric pollution is sufficient to tip the balance of trees already growing in a stressed situation." Similar declines in red spruce have also been observed in the Adirondacks and in New Hampshire.

It is important to note that trees are able to comb water out of the air with their leaves or needles and branches even when it is not raining. Coniferous trees are particularly effective at doing this because of their twiggy branches and multitudinous needles. For example, when a dense fog moves through a spruce forest, small droplets of moisture that are too light to fall to earth are collected on the branches and needles, where they coalesce and grow. When these droplets are heavy enough, they fall to earth like rain. According to Vogelmann, the branches and needles make up large surfaces, up to about 14 million linear yards per acre. Dwight Leedy, a graduate student in botany at Vermont, determined that the forests in the Green Mountains comb at least five inches of water a year from clouds, and Vogelmann reported that "In some areas of these forests, such as the mountain summits where the winds are strong, some thirty inches of water per year could be raked from the clouds and fog by the trees. These amounts of water would, of course, be added to the area's annual precipitation of between forty and sixty inches."

When we last spoke to Vogelmann, in April of 1982, he said that there had also been dramatic declines in sugar maples and beech trees in the Green Mountains. The density of sugar maples dropped 32 percent between 1965 and 1979, while the basal area (the standard of measurement used for all the standing wood bunched together) was down by 15 percent. For beech the density declined by 47 percent and the basal area by 30 percent. "Not only are we losing trees, but we're losing the big ones," Vogelmann said. "The picture is not good."

Siccama and A. H. Johnson of the University of Pennsylvania have also noted an abnormal decrease in growth rates over the past twenty-five years for both pitch pine and loblolly pine in the New Jersey Pine Barrens. The statistical relationship between stream pH and growth rates suggests that acid precipitation may have been a factor.

In an experiment conducted at the Coweeta Hydrologic Laboratory near Franklin, North Carolina, Bruce Haines, Marcia Stefani, and Floyd Hendrix of the University of Georgia subjected eight different plants to artificial rains ranging from pH 2.5 to 0.5 in an effort to determine the threshold for and the symptoms of damage. The species were flowering dogwood, red maple, tulip tree, tanbark oak, northern white pine, pecan, and *Erectites hieracifolia*, a herbaceous weed. Results of the experiment "suggest that a tenfold increase in acidity from pH 3.2 to 2.2 in *a single spring or summer storm* could bring damage or death to mature leaves of dominant flowering plants in the southern Appalachians." (Emphasis added.)

Air pollution is damaging New England in a number of ways. In a paper, "The New England Landscape: Air Pollution Stress and Energy Policy," Dr. F. Herbert Bormann of Yale University writes that the Connecticut–New York City–Philadelphia area has the worst photo-oxidant pollution in the country after the Los Angeles Basin and that as long ago as 1953, Connecticut suffered a $5 million loss in crops from ozone. In Fitchburg, Massachusetts, 66 percent of all recorded ozone values that exceeded the national standard occurred when winds came from the New York City area. Moreover, ozone concentrations sufficient to damage test tobacco plants on Nantucket have been correlated with across-the-ocean winds coming from the New York-to-Washington corridor. The highest levels of cadmium deposition in New England occurred at high elevations in the Green Mountains, and the levels of lead in some rainstorms in north-central New Hampshire have exceeded the U.S. Public Health Drinking Water Standards.

Bormann estimated that between 700 to 1,400 species of plants in New England might be sensitive to sulfur dioxide and/or ozone, often at doses below the ozone standard "and that loss of [plant] productivity can occur without any visible symptoms of injury." New England species known to be sensitive to ozone are white oak, tree-of-heaven, trembling aspen, Concord grape, honey locust, tulip poplar, sycamore, and white ash. Species sensitive to sulfur dioxide are red and white pine, black willow, trembling aspen, large-tooth aspen, low-bush blueberry, staghorn sumac, and yellow and white birch. Yellow and white birch and white ash are suffering decline or dieback in different parts of New England. Elsewhere in the United States, white pine "along the Blue Ridge Parkway in western Virginia have been noted to die following clinical symptoms of photooxidant damage," Bormann wrote, while photooxidant stress has weakened Ponderosa and Jeffrey pine in the mountains of southern California. In the last thirty years they have suffered "an enormous mortality, about three percent per annum, which means that hundreds of thousands of trees have died." Photooxidant damage to both these species has now spread to the entire foothills of the western slope of the Sierra Nevada. The death of the pines is usually attributed to pine beetles that invade the pollution-weakened trees.

Bormann stated, "It does not seem unreasonable to conclude that in the face of current levels of air pollution, large-scale biological changes are occurring in our landscape. Wholesale genetic changes are probably under way. Many sensitive species are probably in a state of decline, with decline most rapid in nodes or areas where local pollutants are laid over the regional patterns. Disease and insect outbreaks on plants may be related to the weakened condition of hosts. The likely effect is a decrease in both plant and animal diversity and fundamental changes in ecosystem function."

In an attempt to evaluate the threat of air-pollution stress to the New England landscape, Bormann describes a series of stages leading to ecosystem collapse.

Stage 0. Pristine ecosystems.

Stage I. Pollution occurs at generally low levels, but the functions of the ecosystem are little affected.

Stage IIA. Some species are subtly and adversely affected by increasing levels of pollutants. For instance, plants may suffer a change in reproductive capacity.

Stage IIB. As pollution levels increase, sensitive species decline. "If a sensitive species is an important member of an ecosystem, its loss can have reverberating effects." Furthermore, animals dependent on sensitive plants may suffer, and the same holds true for plants dependent on animals. "Honeybees are particularly sensitive to air pollution, and deleterious effects on them might interfere with pollination of plants, seed production, and ultimately the reestablishment of vegetation."

Stage IIIA. With even more pollution stress, the basic structure of the forest ecosystem changes as trees, tall shrubs, and then short shrubs and herbs die off in that succession. Runoff increases greatly, erosion increases, and local climatic conditions are changed. The likelihood of fire increases because of the accumulation of highly flammable dead wood.

Stage IIIB. Ecosystems collapse. "Even if the perturbing force is removed, these damaged ecosystems would take centuries or even millennia to achieve predisturbance levels of productivity, structure, and function."

Given these stress stages, where does New England stand now? According to Bormann, "No part of

New England is free from some form of air-pollution stress. Some parts are now in Stage I, but a considerable area is probably well advanced into Stages IIA and IIB. . . . However, time, future emissions, and vulnerability of individual ecosystems are factors that could lead to Stage III responses in a number of areas." The sites to watch are mountain forests, especially those where red spruce is already in decline. "If these declining trends prove to be long lasting and red spruce is severely diminished in importance, a Stage III response is not an impossibility. Such a response might be hastened by wildfire burning over the accumulations of dead wood left after the death of spruce."

In Europe there have been disturbing reports of tree dieback, especially of spruce and fir, in Yugoslavia, Czechoslovakia, Poland, East Germany, West Germany, Austria, Switzerland, and France. In West Germany in particular, there is widespread concern and strong suspicion that acid precipitation is causing dieback on a massive scale. Especially hard hit is the spruce *(Picea abies)*, which is known as the "bread and butter" tree of the forest industry because it makes up half of the trees in the West German forests. In November 1981, the newsmagazine *Der Spiegel* ran a three-part series on acid rain—or *Saurer Regen*, as it is called in German, and the articles created a sensation. Specifically the magazine reported that:

In Baden-Württemberg fir trees on 64,000 hectares of land are in a weakened state, and thousands of spruce trees are diseased.

In North-Rhine–Westphalia spruce stands in 58

percent of the forests investigated are in acute jeopardy.

In the Black Forest near Alpirsbach, almost every fir tree, including those 200 years old, has a thinned-out crown, and the treetops have turned gray. Even young trees are losing their needles prematurely. Karl Scheffold, the forest director, said, "If this continues, the Black Forest will have to be renamed the Gray Forest."

In Hesse 3,000 hectares of forest are already diseased. Spruce and pines show sparse needle development, discoloration, and rapid senescence.

In Bavaria 55,000 hectares of coniferous forest in the Franconian Forest or the edge of the Alps are *am Ende,* "at the end."

Der Spiegel observed:

> This environmental calamity appears to damage German forests more than the sum of all those hazards that have always caused the forest industry trouble: red rot and weevils, forest fires, and foraging animals. For forest experts, the damage reports add up to portray a sickness that some fear will lead to a collapse of the complete ecosystem. . . .
>
> It is now probable that tree damage is not to be traced back to any one single source but evidently is a so-called complex sickness, which nevertheless does appear to have a primary cause.
>
> From month to month scientific indices accumulate which lend more and more weight to a long-discussed hypothesis which is politically very explosive:

—The tree epidemics showing up everywhere probably have a common primary cause: the increasing air pollution over West Germany, caused especially by sulfur dioxide (SO_2) from oil heating, exhausts, and, above all, the smoke stacks of power stations, smelters, and refineries.

—The rapid dying off of conifers, especially endangered by SO_2, is, with increasing sulfur loadings in the air . . . , the beginning of the end for many other kinds of trees.

—Sulfur compounds, which come down on Germany in the form of dust or "acid rain," damage not only the wooded one-third of the country but cause billions of marks of damage on other useful plants and on buildings and increasingly threaten the health of tens of thousands of German citizens.

Der Spiegel went on to note that Professor Peter Schütt, a forest botanist at the University of Munich, warned that if the sulfur dioxide theory is confirmed, West Germany is faced with an "environmental catastrophe of previously unimaginable dimensions." The magazine also quoted Professor Bernhard Ulrich at the University of Göttingen as saying, "The first large forests will soon die in the next five years. They can no longer be rescued."

Dr. Ulrich's words are to be heeded. The director of the Institute for Soil Sciences and Forest Nutrition at Göttingen, he and his colleagues have been studying atmospheric deposition, soil chemistry, and the effects observed on forest vegetation on the poorly buffered Sölling Plateau for fifteen years. Ulrich originally began his research on forest fertilization, but while studying this he became aware of the threat posed by atmospheric deposition of acids. His work did not become generally known in North America until 1981 when a Canadian scientist, Dr. G. H. Tomlinson of the Domtar Research Centre in Senneville, Quebec, visited him in Göttingen and published a special mimeographed report, "Acid Rain and the Forest—The Effect of Aluminum and the German Experience." Dr. Tomlinson summarized the findings of Ulrich and his colleagues as follows:

> The forest foliage and bark capture and oxidize sulfur dioxide present in very low concentration in the atmosphere. The resultant sulfuric acid together with that contained in aerosols, also deposited in the canopy from the atmosphere, are washed from the trees with rain. In the beech forest, as a result, acid reaching the forest floor is annually twice that in the rain entering the canopy. In the spruce forest, where the needles form active sites in the winter as well as summer, the total quantity of acid under the canopy is approximately four times greater than that in the precipitation.
>
> Carbon and nitrogen stores in the soil are increasing as a result of the slow breakdown of humus under the more acid conditions prevailing.
>
> Reactions in the soil, particularly those involving changes in nitrogen, generate further quantities of acid. In the beech forest, the internally generated

acid, together with that reaching the forest floor, is over five times greater than in the precipitation.

During the period in which these ecosystems have been studied, major changes in soil chemistry have occurred involving the release of aluminum into the soil solution as a result of the reaction of the acid with the aluminum silicate in clay.

Necrosis of the fine roots has occurred as a result of the toxic action of inorganic aluminum salts in the mineral soil. The organically complexed aluminum in the humus layer is apparently less toxic, and the fine rootlets in this upper zone allow survival of the trees in a highly stressed condition involving less foliage, dry and brittle crowns, etc.

The condition of the trees is such that their ultimate survival is unlikely and, at least in the case of the beech forest, seedlings will not survive because of aluminum toxicity to their roots.

As with agricultural soils, liming appears to be the only remedial measure to rectify the damage already done to the soil in these areas.

Analysis of soil cores taken from areas in which fir dieback has been observed, and which are outside the main study zone, indicate that aluminum toxicity is involved.

After visiting Ulrich a second time in 1981, Tomlinson published a follow-up report, "Die-Backs of Forests—Continuing Observations." In this paper Tomlinson took note of dieback of spruce in southwest Sweden; dieback of spruce, fir, and beech in the New Forest in England; dieback of red spruce in the Adirondacks, Vermont, and New Hampshire and concluded, "It is vital that, whatever their nature, the underlying causes for this ecological threat, with its esthetic and extremely serious economic implications, be established beyond all question. It is urgent that the problem be given immediate attention."

There is particular concern about potential forest degradation in Canada, where forestry is the single largest industry. The concern is that by the time a 15 to 20 percent loss in forest productivity is proved, it may be too late to save the trees that remain.

Human health can be adversely affected by acid precipitation in a number of ways. We noted earlier that acid precipitation can mobilize toxic metals from bedrock and soils and contaminate fish. Acid precipitation also can leach metals such as copper and lead from water pipes. Reservoir supplies, wells, and springs can be affected. Even the rain that falls from the sky can be dangerous to homeowners who obtain their drinking water from roof catchments. In a study of forty roof-catchment homes in Clarion and Indiana Counties in western Pennsylvania, Dr. William E. Sharpe of Pennsylvania State University found that twenty-eight homes, 70 percent, had airborne lead concentrations that exceeded the safe level. Nine homes had hazardous lead concentrations caused by acid corrosion of the plumbing. "People who rely on roof catchments have a very definite problem," Sharpe said. "There are possible health effects, and there are economic costs. They're going to have to lay out $500 to $1,000 per household to make rudimentary changes. The rural sections of Clarion and Indiana Counties have no public water supplies, and there is an irony here: Deep and surface coal mining has polluted the ground and surface waters so they're unfit to drink, and the people have turned to the sky as a last resort. It's the coal that's mined in the area and shipped to power plants that's coming back to kill their last resort. In Ohio, there are 67,000 rainwater cistern systems. Those people have a problem, and they don't know it."

Aluminum in water is a danger to kidney-dialysis patients. Introduced directly into the bloodstream, aluminum in the dialysate bypasses the normal absorption barriers in the body. In Great Britain and Minnesota, aluminum in the dialysate water showed a significant correlation with the incidence of either dialysis encephalopathy, a brain disease, or fracturing dialysis osteodystrophy, a bone disease. In Minnesota the epidemic of brain disease among dialysis patients was attributed to high levels of aluminum in the tap water, which occurred when the deionizer used to treat the water did not work properly. The source of the aluminum in the tap water was aluminum sulfate used to treat organic contaminants in the water drawn from the Mississippi River. In Britain ten patients in their twenties on home dialysis with aluminum in the water supplies all died within two years.

In recent years investigators have noted high levels of aluminum in the brains of persons afflicted with Alzheimer's disease, a premature progressive memory loss and brain deterioration similar to senility. The cause or causes of Alzheimer's disease, which is irreversible, are not known, but investigators are trying to determine how aluminum might affect brain function.

There is also concern about the inhalation of suspended microparticulates of nitrate and sulfate materials that can penetrate into the lungs. They have been associated with asthma, bronchitis, cancer, and cardiovascular diseases, along with other adverse health effects. "We are concerned now that fine particles are causing widespread health damage in the Northeast," said Dr. Michael Oppenheimer of the Environmental Defense Fund. "Some estimates of mortality associated with these particles exceed the level of tens of thousands of deaths per year." The killer fog that caused an estimated 4,000 deaths in London in December 1952 was associated with short-term acute exposure to the sulfur oxide-particulate complex. In recent years the atmosphere in the western United States has become increasingly polluted, but, said Oppenheimer, "For years asthmatics have been told to go west for their health, and the inhalation of microparticulates can especially affect them as well as old people and children."

In assessing the effect on human health and welfare, the *Northeast Damage Report* states, "Extensive data confirm that sulfates and other fine particles involved in atmospheric deposition are associated with direct effects on human health, from serious respiratory and cardiovascular diseases to death from cancer. Adverse health effects have been identified at levels well below the annual sulfate levels found in the Northeast, sulfate levels that are attributable in large part to long-range transport into the region. The high levels of sulfates in the Northeast definitely contribute to a significant number of premature deaths, and the probability of dying from air pollution-related diseases is significantly higher in the Northeast."

The National Wildlife Federation has raised the concern that acid precipitation may promote the accelerated leaching, mobilization, and enhanced biological availability of toxic metals and chemicals in solid waste and hazardous waste landfills, which number in the tens of thousands in the United States. This potential problem really has not been addressed.

In Stockholm, Professor Torbjörn Westermark, a nuclear chemist at the Royal Institute of Technology, has applied to Swedish environmental authorities for funds to study the effect of acid precipitation on metals, including radium. Westermark, who was one of the first scientists to sound the alarm on mercury contamination of fish in the 1960s, is concerned that acidification might release decayed products of uranium from bedrock. If and when he gets the funding, and this seems likely, he will begin to measure the radium content in the shells of freshwater clams and mussels, fishbones, otter bones, and the bones of fish-eating birds. He also plans to test human bones because radium accumulates in the bone close to the blood-making organs of the body, and he is concerned about the risk of leukemia. For comparison, Westermark plans to look at the contents of bones collected in the past. "We can start to look in the Natural History Museum in

Stockholm," he said. "It began to collect its samples in 1830, so we hope to have measurements from shells, bird skeletons, etc., from before the Industrial Revolution. What we applied for was funding for investigation of heavy metal leach-out. After we get a picture of the historical aspect, we can then do a forecast for the next ten to twenty years."

Westermark added, "We haven't yet applied for testing human bone, but they won't object if we include this. Archeologists have located bones from the thirteenth century, which should provide a historical comparison. If one finds an increase of radium in human bones by, say, a factor of ten, then it would be worthwhile to launch a major project. The risk level of leukemia is low compared with smoking, air pollution, etc., [but] one point would be to investigate the population living on primary rock, granite and gneiss, which has no buffer capacity, no resistance to acid rain. Limestone, on which most Western Europeans live, neutralizes acid rain."

Acid precipitation is eating into metals, marble, limestone, and calcareous sandstones, and it is accelerating the degradation of statues, buildings, and monuments in the United States and abroad. Some of those affected in this country include the Statue of Liberty, the Field Museum in Chicago, the Washington Monument, and the Capitol. "The east side of the Capitol is white Lee marble from Lee, Massachusetts," said Dr. Erhard Winkler, professor of earth sciences at the University of Notre Dame and a consultant to the National Bureau of Standards and the National Park Service. "There are craters one-quarter inch or more in it. It looks like shrapnel has hit it." Because of acid precipitation, the hard minerals in the marble have changed to talc. (It is sometimes reported that acid precipitation has damaged Cleopatra's Needle, the obelisk in New York's Central Park. Not so, says Winkler who attributes the flaking to infiltration of salts when the obelisk was in Egypt and its subsequent removal to the relatively humid climate of New York. Nowadays American museums make it a standard practice to place all stone monuments arriving from desert regions in distilled water to leach salts.)

Acid degradation would cost the State of Massachusetts almost $8 million to refurbish bronze statues alone, and annual maintenance would cost between $290,000 and $370,000. In 1981 the United States/Canada Work Group on Transboundary Air Pollution recognized that "the architectural and sculptural expressions of our two heritages are a nonrenewable resource of the most precious sort. Historic structures and monumental statuary represent the most visible aspects of historical and cultural evolution." Laws in both countries mandate the preservation of historic objects, but nothing is being done to enforce these laws.

Acid precipitation also causes the deterioration of mortar, concrete, paper, paint, leather, and textiles. Back in 1941 there were a number of episodes in Washington, D.C., in which women complained of holes and runs in silk and nylon stockings. Local smoke-control officials discovered that the stockings had been ruined only on windy days and that they contained particles of sulfuric acid. Officials blamed improper fuel combustion, and a photograph in *Life* showed an inspector holding stockings over a chimney with, ironically, the Washington Monument in the background.

Dr. Norbert Baer of the Conservation Center of the Institute of Fine Arts at New York University is conducting an unusual but potentially significant study for the U.S. Environmental Protection Agency. He is assessing the effects of acid precipitation on gravestones in more than 100 national cemeteries throughout the country. Veterans first became eligible to receive gravestones in 1873, and inasmuch as the stones are either marble or granite and have come from only two or three quarries, they provide ready-made sets of chemically uniform indicators that have inadvertently recorded the effects of corrosion at each site over precise periods of time.

In 1981 the National Wildlife Federation reported that acid-deposition damage to automobiles had increased in the past twenty years. Acids either etch the surface finish or leach pigments from the paint surface. After acids have penetrated the finish, it is costly to refinish the car because the old finish has to be stripped down to bare metal. The federation said that in 1980 the highest incidence of reported auto damage from acid precipitation occurred in Florida, New Jersey, New York, and Pennsylvania. In 1978 several hundred cars in central Pennsylvania were scarred by raindrop-shaped polka dots following a rainstorm with a pH of 2.3. Sulfuric acid leached the yellow from a lime-green Maverick, leaving the car spattered with baby blue polka dots. It is a sign of the times that the Exxon station off the Route 3 rotary in Chelmsford, Massachusetts, advertises, "Wash the acid-rain dust off your car."

Air pollution can affect weather and climate. Bormann has already pointed out that Stage IIIA of a forest's ecosystem collapse would change local climatic conditions, and there are signs that changes can occur on a larger scale. Particulate matter reduces the penetration of sunlight, prompting Dr. Michael Oppenheimer to say, "I think it likely that this will have some effect on climate in the Northeast and perhaps elsewhere."

Reduction of sunlight also influences plant photosynthesis and alterations in the length of the growing season, both of which can have an impact on the productivity of farmlands and forests—especially in the Northeast, which has a high relative humidity that exacerbates the problem. Sulfate particles in particular can scatter light in the atmosphere and reduce visibility. Between 1953 and 1972, visibility decreased 10 to 40 percent in rural areas in the Northeast, where visibility now averages less than eight miles. "Visibility is especially important in recreational areas of New England and New York because of its aesthetic values to the tourist industry," the *Northeast Damage Report* stated. "Tourism is vital to New England's economy, particularly in the rural and mountainous areas. A visitor who has driven for several hours and then hiked for several more only to find that there is no view from the top of the mountain is not likely to return."

This is happening elsewhere in the country. Man-made haze often obliterates the natural blue haze of the Great Smoky Mountains, and, according to Dr. Hugh Spencer of the University of Louisville, visibility in the Ohio River Valley has decreased from an average of ten to twelve miles in 1939 to four miles at present. Oppenheimer said, "On certain days the air pollution from Los Angeles blows across the Southwest. Add that pollution drifting out of the L.A. Basin to that from the large power plants and smelters in the Southwest, and some of our best views in the West, such as the view of the surrounding ridges from within the Bryce Canyon National Park, are reduced. The haziness is caused by the microparticulates associated with acid precipitation. Congress was specifically concerned about this when it amended the Clean Air Act."

Scientists who have studied visibility records for airports in the eastern United States report that in the early 1950s summer had the best visibility but it now has the worst. From 1948 to 1974, summer haze increased by more than 100 percent in the central eastern states, by 50 to 70 percent in the Midwest and eastern Sunbelt states, and by 10 to 20 percent in New England. As the United States/Canada Work Group of scientists reported in their *Phase II Interim Working Paper*, October 1981, "Very close parallels have been noted between the geographical/seasonal features of airport-visibility trends, and the geographical/seasonal features of trends in atmospheric sulphate concentrations, sulphur dioxide emissions, and coal-use patterns. These parallels provide strong circumstantial evidence that the historical visibility changes in the East were caused, at least in part, by trends in sulphate concentrations and sulphur dioxide emissions." The Work Group added, "Episodes of regional scale haziness have been observed in the eastern United States and Canada. Examination of airport data, pollution measurements, and satellite photography indicates that these haze air masses move across eastern North America in the manner of high pressure systems, causing significant visibility reductions in areas with little or no air pollutant emissions."

After observing a thunderstorm in the Hudson River Highlands in the summer of 1980, the junior author of this book wondered if acid precipitation might have an effect on the incidence and intensity of thunderstorms and the frequency, duration, and direction of lightning. Many of the strikes he observed were striking down into shallow coves, not the adjacent higher ground and hills. Was the rain, presumably acid, acting as an electrolyte, a conductor, so the lightning strikes followed its downward path? Could an acid thunderstorm intensify by acting synergistically upon itself? Later we wondered if acid-induced lightning could have caused the devastating Con Edison power blackout of July 13, 1977, which plunged Westchester County and New York City into darkness and prompted an orgy of looting. The initial cause of the loss of power was a "freak" lightning strike at 8:37 P.M. that took out two major feeders between the Buchanan and Millwood substations in northern Westchester. The odds of this happening are considered very remote, but at 8:55 a second "freak" stroke in the same area took out two more feeders, and the statistical chances of this happening are considered all but impossible. Unfortunately we were unable to learn the pH of the rain in that thunderstorm because, we discovered, there were no sampling stations in the region in 1977 that measured the pH of precipitation. We are aware, of course, that we may be indulging in speculations for which we may be derided, but these are speculations that need to be addressed. Certainly papers and reports in the meteorological literature demonstrate that air pollution can increase the frequency, intensity, and duration of thunderstorms or otherwise alter weather.

One example is the so-called "La Porte Weather Anomaly" investigated by Stanley A. Changnon, a climatologist with the Illinois State Water Survey. This involved the city of La Porte, in Indiana, to the east of major steel mills and other heavy industry in the Chicago-Gary area. At the 1968 meeting of the American Meteorological Society, Changnon reported that between 1951 and 1965 La Porte experienced 31 percent more precipitation, 38 percent more thunderstorms, and 246 percent more hail days than did nearby weather stations in Illinois, Indiana, and Michigan. He correlated the increased precipitation in La Porte with increased iron and steel production in Chicago and Gary. To Changnon it appeared that industrial pollutants were serving as nuclei that triggered precipitation in much the same way that silver iodide crystals are used to seed clouds. La Porte's freaky weather stopped in the 1960s, possibly because the general circulation pattern changed and shifted the anomaly into Lake Michigan, where it could not be readily observed.

Some climatologists attacked the validity of the La Porte anomaly mainly because it was based on only circumstantial evidence. In 1973 Changnon and F. A. Huff reported in the *Bulletin of the American Meteorological Society* that study of historical weather records in and around Indianapolis, Cleveland, Washington, Chicago, Tulsa, Houston, New Orleans, and St. Louis indicated that increased precipitation is related to city size, industrial nuclei generation, and urban thermal effects.

The research in the St. Louis area was only part of a five-year project known as METROMEX, an acronym for Metropolitan Meteorological Experiment. Conducted by a team of scientists, the METROMEX report was published in 1977, and the findings were startling. Briefly put, the St. Louis urban area altered atmospheric processes in a variety of ways. East of St. Louis in Illinois, summer rainfall was found to be 5 to 30 percent greater than elsewhere in the area. The acidity of precipitation was not dealt with at any length, probably because the study was done before it became a major issue, but the report did note that the East St. Louis and Granite City area experienced, "on the average, quite acid rainfall." Air heated and polluted by St. Louis moved high enough into the atmosphere to affect clouds, and a light cumulus formation could conceivably turn into a heavy thunderstorm. Locally triggered storms occasionally initiated other local storms, which tended to merge to produce heavier rains, stronger winds, and hail as they moved into Illinois. Indeed, the frequencies of thunderstorms east of St. Louis were 45 percent greater than elsewhere. Moreover, the duration of thunder periods was 56 percent greater, and the periods of lightning frequency were 83 percent greater. The area east of St. Louis also had more hail days, more hailstorms, and greater hail intensity in the number and sizes of the hailstones. The heavier rains in Illinois reduced highway visibility and made roads slippery—conditions that could help cause accidents, injuries, and deaths. How many accidents were attributable to the urban-affected severe weather was not known, but the impact was "likely of major significance." Moreover, downdrafts from thunderstorms were especially threatening to large jets on landing. "The crash of a commercial airliner on its landing approach at St. Louis in 1973 appeared to occur in the gust front generated by downdrafts of an urban-affected thunderstorm."

There were other impacts. A 30 to 50 percent increase in rain east of St. Louis often caused sewage-treatment plants to bypass the surcharge, which added to stream and river pollution. Floodplain communities sometimes suffered sewer line breaks because of shifting caused by groundwater fluctuations. "The heavy rain anomaly has helped cause many communities east of St. Louis to have difficulty in meeting state and federal standards for waste treatment." The heavy storms also helped increase local flooding by 100 percent, and proposed flood-control projects were going to cost some $73 million. There was also a 34 percent increase in erosion, which further degraded local water quality.

The added summer rain increased the average farm income by $832 annually, mostly because of added soybean yields, but the effects of acid rain or pollution deposited by the rain could not be ascertained. "The added zinc deposited in the area should have a beneficial effect on the local horseradish production since zinc aids in its growth." Increased hail damage to crops, however, could lead insurance companies to raise their rates. Air-pollution injury to trees and other plants was a problem in the St. Louis urban area, but it was not clear how much of the injury could be attributed to pollutants added by the increased rain.

There was an increase in power outages caused by lightning strikes, particularly in the suburbs east of St. Louis, that were both inconvenient and costly to the Illinois Power Company, the public, and business and industry.

The study showed that heavier rain and poor visibility resulted in increased crime, which leads to added costs in law enforcement. Thus taxes for law enforcement and sewage systems should increase, but at the same time income from property taxes was likely to decrease because, in general, air pollution reduces median property values. Put together, all these factors formed a sequence of events that acted to help cause or reinforce migration out of the city by the more wealthy population to suburbs beyond the affected area. This in turn resulted in relatively more low-income residents in the urban area and less taxable income for affected urban governments. "In a sense, the urban-produced rain anomaly helps produce a series of events that lead to urban decay." Finally the METROMEX report posed a key question that it did not answer. Could the potential combined effects from megalopolises "be additive and trigger climatic changes" on a much larger scale?

Above and beyond local changes, scientists have identified two long-term changes in the global atmosphere brought about by air pollution, in addition to acid precipitation. The burning of fossil fuels is also raising the atmospheric concentrations of carbon dioxide. Should this continue, it could cause a warming of the earth—the so-called greenhouse effect—before the year 2050 that would be sufficient to alter precipitation patterns and induce climatic changes on a global scale, leading to widespread ecological, agricultural, social, and economic turmoil. The second long-term change is the depletion of the stratospheric ozone layer that shields the earth from damaging ultraviolet radiation. It is believed that various emissions, notably chlorofluorocarbons used as aerosol propellants, refrigerants, blowing agents in foam production and solvents, are depleting the protective ozone layer, and the resultant increase in ultraviolet radiation would be likely to increase the incidence of skin cancer and to cause damage to agriculture.

5

The Politicians Stall

What is being done to curtail acid precipitation in North America? A fair amount, in Canada. The single largest source of sulfur-dioxide emissions on this continent is in Sudbury, Ontario. It is the International Nickel Company's nearly quarter-mile-high superstack, the tallest in the world. This stack used to spew 5,000 tons of sulfur dioxide each day into the atmosphere, and scientists working on acid precipitation half-jokingly began to use the term *Sudbury* as a unit of measure, as in "Ohio is two Sudburys" or "Indiana is one Sudbury." Now the Sudbury superstack is down to 2,500 tons of sulfur dioxide a day, and the company is under provincial orders to cut emissions to 1,950 tons a day by 1983 and make further reductions thereafter to the lowest feasible level. Until 1980 the provincial governments set the emission standards for sulfur and nitrogen oxides; but Canada amended the law, giving Parliament the authority to control sources that contribute to pollution "across national boundaries."

But it is absolutely impossible for Canada to go it alone. Two-thirds of the sulfuric acid that falls on Canada originates in the United States. In August of 1980 Canada and the United States signed a Memorandum of Intent concerning transboundary air pollution in which both nations agreed "to combat transboundary air pollution in keeping with their existing international rights, obligations, commitments and cooperative practices," including those set forth in the 1909 Boundary Waters Treaty and the 1978 Great Lakes Water Quality Agreement. Canada and the United States also agreed to establish a Coordinating Committee to begin discussions, and both nations pledged to "develop domestic air-pollution control policies and strategies, and as necessary and appropriate, seek legislative or other support to give effect to them."

Although studies have been going on since then, there is no indication that the United States, at least under the Reagan Administration, is prepared to carry through on the Memorandum of Intent in timely fashion. Perhaps the clearest insight into the administration's thinking was given before Reagan came to power, in 1980, by David Stockman, now the Director of the Office of Management and Budget. At a Washington meeting of the National Association of Manufacturers, Stockman said he was a "self-avowed heretic" who did not belong to the "choir of the faithful committed to issuing melodious harmonies to the tenets of orthodoxy regarding the Clean Air Act." Stockman said almost every existing standard was already "far too stringent relative to both what economic and public policy and the medical evidence would suggest."

Addressing himself directly to acid rain, Stockman went on to say:

> I don't know how closely any of you follow the national papers, but if you read the [*Washington*] *Star*, the *Post*, and the *New York Times* you find out that somebody's orchestrating a pretty careful strategy because every other day there's a new article about the acid-rain problem. And it's written by reporters who know not a damned thing, and you'll excuse my language, about pollution, the techniques of pollution, the chemistry of pollution. And they're writing such preposterous and absurd things that what it's doing is creating an intellectual climate, an attitudinal climate, that will probably cause EPA or the Congress to lurch forward into an acid-rain program that's based on nothing more substantial than the tall-pipe standards were in 1970.
>
> I kept reading these stories that there are 170 lakes dead in New York that will no longer carry any fish or aquatic life. And it occurred to me to ask the question . . . well how much are the fish worth in these 170 lakes that account for four percent of the lake area of New York? And does it make sense to spend billions of dollars controlling emissions from sources in Ohio and elsewhere if you're talking about very marginal volume of dollar value, either in recreational terms or commercial terms?

After Stockman finished, an NAM spokesman said he found it "encouraging to know that somebody who thinks like that is still in Washington and has something to say." Stockman now has much to say about any approach to acid rain. As Director of the Office of Management and Budget, he has oversight of all federal environmental regulations.

After Reagan won the presidency, Prime Minister Pierre Trudeau of Canada said that he planned to discuss acid precipitation with the new President at the first meeting between the two. They met and talked in Ottawa in March 1981, but little came of their discussion. A month later, Robin Porter, the State Department's specialist on pollution problems with Canada, said that any treaty with Canada on transboundary air pollution was "at least three or four years away." Canadian officials said that any such delay was unacceptable.

Several weeks later, the administration further angered Canadians when it failed to send its official representative—Frederic N. Khedouri, Stockman's Associate Director for Natural Resources, Energy and Science at the OMB—to an international conference on acid rain in Buffalo. Among those stood up by Washington was Dr. Mark MacGuigan, Canada's Secretary of State for External Affairs, who pointedly told the conference:

> To . . . those who propound the view that economic and energy considerations make significant controls unfeasible, I would submit that significant emissions reductions, if wisely applied, need not

detract from economic and energy goals. Nor should the legitimate costs of production be passed off to another party, in this case another country. This is spurious in economic terms and irresponsible in the spirit of international legal considerations. . . .

Acid rain is a serious bilateral issue because Canadians perceive that further delay in tackling the burgeoning threat of acid rain can result in further incalculable damage. Such delays would be particularly repugnant to Canadians if they were solely the result of narrow, vested interests. . . .

It was an international arbitration in the 1930s between Canada and the United States that provided what is still the clearest statement of the international law relating to air pollution. At the conclusion of the Trail Smelter Arbitration, in which Canada had previously accepted liability for damage caused [to farmers] in the State of Washington by fumes from a smelter in British Columbia, the arbitral tribunal stated that "no state has the right to permit the use of its territory in such a manner as to cause injury by fumes in or to the territory of another, or the properties of persons therein."

I am certain that all responsible Americans accept that the rule of law should guide their relations with other countries as well as their internal activities. I am also certain that responsible Americans recognize that our mutual obligations must be met by dealing with the causes of acid rain to prevent further damage rather than concentrating on remedies for damage after it has occurred.

Talks have continued between the two countries, and at a meeting in Washington in March 1982, Canada agreed to achieve a 50 percent reduction in sulfur-dioxide emissions by 1990 if the United States agreed to do the same. The Reagan Administration failed to respond in kind, and an exasperated Canadian official complained that the United States had been "less than forthcoming with suggestions" since signing the Memorandum of Intent.

Stalling is the name of the game with the Reagan Administration and its allies in industry who are determined to keep on polluting because it is profitable for them to do so. If that means inflicting possibly irreversible damage on others, so be it. Big bucks are involved, and as Governor James Rhodes of Ohio—the single biggest coal-burning state in the country—once remarked, "Those Scandinavians and Canadians—when you corner them, they say they *think* this or that. Well, we want them to think—think they're playing with big money."

On the state level, the Reagan Administration has had no bigger ally than Governor Rhodes, who once maintained that Ohio was no more to blame for acid rain than Florida is for hurricanes. Under Rhodes's leadership, Ohio has doggedly fought against curbing emissions. In testimony before Congress in 1980, James F. McAvoy, then director of Ohio's Environmental Protection Agency, refused to concede that acid precipitation was "a very serious problem." This prompted Rep. Marc L. Marks of Pennsylvania to remark, "You may be the only person here who won't." When Rep. Doug Walgren of Pennsylvania asked McAvoy if he agreed with a U.S.–Canadian research group report that two-thirds of all U.S. sulfur-dioxide emissions come from electric-generating plants, McAvoy answered, "I do not feel knowledgeable, nor do I think anybody is knowledgeable enough, to agree or disagree." The thrust of McAvoy's testimony was to stall. He testified, "Despite the reported effects of acid rain on the environment we cannot afford to overreact to preliminary data, especially in light of our grave energy needs today. By U.S. EPA's own admission, the question of acid rainfall is one that should receive close attention in terms of research and development because of the current lack of knowledge of its origins and effects. We are aware that both the U.S. EPA and the [Carter] White House have stated that it will take at least ten years to accurately determine the extent, effects, sources, and controls of this phenomenon. In our opinion, the ten-year figure may be overly optimistic based upon the inconclusive results of the extensive studies of the same phenomenon during the last decade on the European continent."

As the director of Ohio's EPA, McAvoy consistently sided with industry against what he called "no-growth extremists," "environmental extremists," and "radical regulators," and a former colleague said that recommendations for legal action against polluters had a way of "disappearing" after they reached McAvoy's desk. In a speech before the Capital City Young Republicans in Columbus, McAvoy said, "We refuse to be coerced and cowed by a small group of coercive utopians who are hell-bent on pursuing a zero-pollution, no-growth policy." He added that "the new Orwellian federal bureaucracy has come to be dominated by a small group of 'Big Brothers' who feel that regulation, and plenty of it, is the best way to keep individual Americans and of course 'big bad business' walking the straight and narrow road to Utopia." He added, "And Utopia for this group can't be called anything other than socialism." In reporting these comments, the Columbus *Dispatch* observed, "McAvoy, however, plays first trumpet rather than being the composer and his policies have echoed Rhodes in combating the U.S. Clean Air Act and U.S. EPA over the burning of high-sulfur Ohio coal, giving general support for nuclear power (as long as the wastes go elsewhere) and urging strong sensitivity to the problems of industry and labor in complying with antipollution laws."

After Reagan won the presidency, Governor Rhodes urged that McAvoy be appointed administrator of the U.S. EPA, but when that post went to Anne Gorsuch, a James Watt protégé, Rhodes backed McAvoy for one of the three positions on the Council on Environmental Quality (CEQ), which had been given the job of coordinating the administration's policy on acid precipitation. Reagan nominated McAvoy for the CEQ, but in September 1981 he withdrew it after it became known that McAvoy had taken "poetic license," as McAvoy himself put it, with his résumé. Governor Rhodes didn't quite see it that way, declaring that "Jim McAvoy is the victim of a campaign by environmental groups and former Carter appointees who simply refuse to admit that industrial states should have a say in what goes on in Washington. McAvoy was asked to go to Washington and join the new administration as a voice of the industrial states. He was there to fight for the rights of steelworkers, coal miners, unemployed auto workers, and other industrial workers whose jobs are threatened daily by bureaucrats in the U.S. EPA and other regulatory agencies." The governor concluded by saying that "Jim McAvoy is a talented and intelligent man, and I am certain his services will be in demand."

They certainly were. It turned out that when his nomination was withdrawn McAvoy was already employed as a senior staff member at the CEQ at a salary of $50,112.50 a year, not much less than the $52,750 he would have received as a council member. Moreover, McAvoy said he would stay on as a senior staff member to continue his work on acid rain, the Clean Air Act, and other environmental issues. In the end, Rhodes's first trumpet has wound up in a position of considerable influence, and the Reagan Administration didn't have to go through a potentially messy confirmation fight in the Senate to get him there.

At about the same time that President Reagan withdrew McAvoy's nomination for the CEQ, Governor Rhodes received the report of the State of Ohio Scientific Advisory Task Force on Acid Rain. Rhodes had formed the task force in March of 1980 because he was becoming sensitive to the bad press Ohio was getting. Rhodes told the task force to "let the chips fall where they may," but only after he had blasted "no-growth environmentalists" who had "latched on to acid rain as a rallying cry for a new wave of environmental hysteria."

The members of the task force included representatives of the utility and coal-mining interests, the heads of various state agencies, and Dr. Thomas A. Seliga, director of the Atmospheric Science Program at Ohio State University, who had been instrumental in holding the first national conference on acid rain in 1975. For eighteen months the task force conducted public seminars to educate task-force members as well as the general public about the problem, and task-force members also independently reviewed current literature and attended regional and national meetings on acid precipitation.

Given Ohio's previous intransigence, the report laid it on the line to Governor Rhodes. It recommended the following:

> A. The Task Force believes the State of Ohio must address itself to acid rain specifically and air pollution generally in a forthright manner. Ohio EPA should develop the internal capability to address each of the major air-pollution issues confronting Ohio. Additional university programs in air pollution studies should be encouraged. A system of support should be established to ensure that Ohio universities and private research foundations can take advantage of all available resources to perform air-pollution research. The Ohio Air Quality Development Authority should be considered as one possible source of support for acid-rain-related research in Ohio.
>
> B. The current political concern about acid rain could

result in the adoption of premature regulatory strategies which would limit emissions of SO_x and NO_x more than at present. The Task Force believes the State of Ohio and its industries must be prepared to deal with the consequences of such actions. A capability must be developed to fully examine the socioeconomic impacts of possible acid-rain-control strategies. Several different strategies should be examined with options ranging from no change to complete control (using the latest proven and cost-effective technology). It is believed that if Ohio moves ahead to meet its commitments to air-quality goals, the reduced emissions will help to alleviate the potential for acid-rain formation.

C. Because of the interstate and international aspects of the acid-rain issue, the Task Force recommends creation of an *ad hoc* panel of representatives of all concerned states and Canadian provinces as well as the United States and Canadian federal governments to draft the framework of a proposed program which might be formally enacted in the two countries. All members of the *ad hoc* panel should have equal status.

D. Ohio has been criticized for the failure to obtain federal approval for the state sulfur-dioxide-control program. The state's sulfur-dioxide plan has been awaiting federal approval since September of 1979. The Task Force believes that Ohio EPA should work with the U.S. EPA to reach an agreement regarding the appropriate sulfur-dioxide-control plan for Ohio.

E. The Task Force recommends that all high-sulfur coal burned in Ohio be washed as an initial control program. This would result in a reduced atmospheric loading of from 500,000 to 700,000 tons per year (or 16% to 25%) and would have an immediate impact in reducing the potential for acid rain which might be attributed to this state.

F. The Task Force believes that proposals for technologies other than coal washing should not be considered without thorough social, economic, and environmental impact studies.

G. The Task Force supports the requirement that all new major coal-fired electric-generating facilities be designed and operated to meet federal New Source Performance Standards. Over the long term, the replacement of older more-polluting facilities by new, cleaner plants will result in reduced emissions and cleaner air.

At this writing, Ohio has done next to nothing to implement the recommendations of the task force, which is, perhaps, to be expected. What has not been expected, however, is the hypocritical flip-flop of the administration of Governor Hugh Carey in New York. No state has suffered more documented damage from acid precipitation than has New York.

A brief bit of background is necessary. During the past couple of years, New York State has pushed for use of coal rather than oil. Given the uncertainties of oil supplies and costs, this makes sense, provided that the emissions from coal, a dirtier fuel, are strictly controlled. By 1994, twenty-one power plants in the state producing a total of 6,000 megawatts of electricity are expected to be burning coal instead of oil. Although the strictest controls on coal will still result in great savings to rate payers, the utilities are out to maximize their profits, and they have clout in Albany.

In the fall of 1981, the state's Department of Environmental Conservation (DEC) completed a study, *Cumulative Environmental Impacts of Coal Conversion*, for the New York State Energy and Research Development Authority (NYSERDA). Instead of publishing the study as received, NYSERDA disemboweled it so that any adverse impacts that coal would have on the environment of New York, adjacent states, or Canada were eliminated or minimized. Here, by way of brief example, is one excerpt from the DEC study and that published by NYSERDA for public consumption. The italicized text in the DEC excerpt shows what NYSERDA deleted.

What the suppressed DEC study stated:

FINDINGS ON ECOLOGICAL IMPACTS

• *The transportation and deposition of pollutants from air masses containing significant concentrations of combustion-related substances and their impact on the environment are of great concern*

What NYSERDA published:

FINDINGS ON ECOLOGICAL IMPACTS

• As determined from emissions modeling (discussed in Subchapter 4a), the concentrations of sulfur oxides (SO_2) from the proposed coal-fired power plants in New York by

in the management of New York's ecosystems. From coal-fired power plants, the primary pollutants of sulfur and nitrogen oxides and the secondary pollutants which form from these gases during air-borne transport may have serious effects on ecosystems.

• As determined from emissions modeling (discussed in Subchapter 4a), the primary concentrations of sulfur oxides (SO_2) from the proposed additional coal-fired plants in New York by 1994 are likely to have little or no direct effect upon ecosystems *except for power plants in close proximity, where under certain "worst case" situations existing ambient air quality nearly exceeds allowable standards and normally allowable primary emissions may overlap. More rigorous emissions control or fuel limitations can minimize such impacts to ecosystems.*

• Emissions modeling shows that transport and deposition of acid precipitation will occur from the proposed additional coal-fired power plants in New York (Subchapter 4c discusses the relative extent of acid precipitation from various emission scenarios). Amounts of deposition from such plants falling upon sensitive areas of New York, like the Adiron- 1994 are likely to have little or no direct effect upon ecosystems.

• Emissions modeling shows that transport and deposition of acid precipitation would occur from the proposed coal-fired plants in New York (Subchapter 4c discusses the relative extent of acid precipitation from various emission scenarios). Amounts of deposition from such plants falling upon sensitive areas of New York, like the Adirondacks, will be small compared to other out-of-state* sources.

*Note NYSERDA's added little touch of "out-of-state."

dacks, will be small compared to other sources, *but it will add slightly to the already serious effects upon both terrestrial and aquatic ecosystems.*

In 1981 the DEC held a hearing on the first coal-conversion application in the state. Orange and Rockland Utilities wanted to convert two oil-burning units, Lovett Units #4 and #5, to 1.5 sulfur coal without controls. The Environmental Defense Fund, the Hudson River Fishermen's Association, and Scenic Hudson intervened in the case. The Lovett units are on the west bank of the Hudson adjacent to the Palisades Interstate Park and the sensitive hills of the Hudson Highlands. The DEC staff, notably the Bureau of Energy, also put up a strong case against any increase in emissions, and the DEC's initial brief noted that "The record shows that emissions from the Lovett facility will contribute to acid precipitation in such vulnerable ecosystems as the Adirondack and Catskill Mountain Ranges, and may have adverse impacts on sensitive areas in the Hudson Highlands. In fact, Applicant's DEIS [Draft Environmental Impact Statement] acknowledges that the Lovett conversion will result in increased acid deposition in eastern North America. The DEIS predicts an increase in SO_2 emissions of approximately 24,000 tons per year with coal use for a total emission of approximately 28,600 tons per year. Witness Lipfert also calculates an SO_2 increase of 21,000–27,000 tons per year from Lovett. DEC Witness Horn has estimated that SO_2 emissions from Lovett will comprise 25 percent of the additional 100,000 tons per year of SO_2 emissions from coal conversions predicted by the New York Power Pool.

"Staff Witness Horn and Intervenor Hendrey have provided comprehensive descriptions of the adverse ecological impacts associated with acid deposition. In addition, EDF Witness Hendrey demonstrated, on the basis of field studies he conducted in the Hudson Highlands, that acid-sensitive areas exist within thirty kilometers of the Lovett facility. These were shown to be precisely the areas which are predicted to receive the highest amounts of acid deposition from the Lovett facility. Mr. Hendrey predicts that increases in acid deposition in the sensitive Hudson Highlands watersheds are likely to result in adverse ecosystems impacts associated with increased acidity."

In April of 1982, Commissioner Robert Flacke of the DEC announced his decision. Flacke did not go along with O & R's blatant request to burn 1.5 percent sulfur coal; instead he offered the utility different conversion options: conversion with .7 percent or less sulfur coal; conversion with 3.8 percent or less sulfur coal and full scrubbing to give the equivalent of .4 percent sulfur coal; or conversion of only one unit with 1.0 percent sulfur coal, with monitoring to determine the advisability of the later conversion of the second unit. Put in plain English, Flacke's options allow O & R to increase SO_2 emissions by up to 15,000 tons a year—a threefold jump, and a disaster for the Hudson Highlands.

A small note to the reader about the Hudson Highlands. Historically, culturally, and aesthetically, the Hudson Highlands are of great significance to the nation. The United States was born there. During the American Revolution, the British grand strategy was to seize the Hudson Valley, divide the rebellious colonies in two, and then dismember them at will. George Washington's counterstrategy was to fortify the Highlands at West Point and other locations. The British, who were never able to seize control after the defeat at Saratoga upriver and the later discovery of Benedict Arnold's treachery, thus lost the war. The Highlands also served as inspiration to Washington Irving and the artists of the Hudson River School who gave birth to American landscape painting. The successful fight against Consolidated Edison's proposed pumped-storage plant at Storm King Mountain at the northern gate of the Highlands helped spark the environmental movement of the 1970s. In an eloquent passage published in 1838, William H. Collyer wrote of the Highlands, ". . . they rise to a height of from twelve to fifteen hundred feet in bold and rocky precipices that would seem to defy the labour of man to surmount, and where the eagle builds her eyrie, and the hawk raises her callow brood, fearless of the stratagems of the spoiler. The mighty river, pent within a narrow channel, struggles around the base of the hills, and lying deep in the shadows of the mountains, often appears like some dark lake shut out from the world. Our country offers no scenes more grand and sublime, and we may well doubt whether the far-famed Rhine, the cherished theme of every European tourist, can exceed in those attributes, the one we are now voyaging upon. There is here, it is true, no classic spot to recall the days of the Caesars, 'no castled crag of Drachenfels' encircled with its thousand years of romance, as on the German river; but to an American it is a land of enduring interest; every spot of this mountain pass, whereon man could gain a foot hold, is interwoven with the history of a struggle that gave birth to a mighty empire, and, in coming ages, may we not suppose that it will be looked upon with some such feeling of veneration as with which we of the present day now contemplate the pass of Thermopylae or the plains of Marathon?"

6

Industry Arguments

For all the help it has had from government, polluting industry takes no chances. The utility and the coal companies have even attempted to take the offensive on acid precipitation by dispensing propaganda designed to mislead both the public and the politicians they do not control. Distortion is common. Take the incident that occurred when F. Peter Fairchild, coauthor of the *Northeast Damage Report*, attended a conference in Chicago on acid precipitation. Fairchild was startled to hear a representative of the Peabody Coal Company read from the *Northeast Damage Report* which had just been published. Fairchild says, "The guy took passages out of my report almost verbatim. And then, obviously, he had people take those statements and those results and try to find other results that would conflict with them or at least confuse them. It was like reading my report with some extra annotated footnotes to confuse the issue. He quoted from another author as contradicting some of the things I said in my report, and the author just happened to be in the room, and he just jumped up and said, 'I can't stand it anymore. You're twisting everything I said.' "

Industry's arguments are often absurd, but there are some seemingly cogent points that need to be answered here so that no one who hears them is gulled. For example:

• *It is unclear whether precipitation is becoming more acid in the East.*

That is correct, but there are large areas of the United States and Canada that cannot endure anywhere near the *present* levels of acidity without suffering further damage.

• *Fish in Florida lakes with a low pH "show no sign of dying."*

That is correct, but what is not said is that the fish are stunted and emaciated.

• *The "three lakes in the Adirondacks" argument.*

This has been a favorite of Dr. Ralph Perhac of the Electric Power Research Institute. On February 27, 1980, Perhac appeared before the House Subcommittee on Oversight and Investigations of the Committee on Interstate and Foreign Commerce and testified on EPRI's lake-acidification study, saying "we have found three lakes in the Adirondack Mountains of New York State which have very different acidities, yet these lakes lie within a few miles of each other and chemistry of the rainfall is the same at all three. Obviously some factor other than precipitation is responsible for the acidity." On March 19, 1980, Perhac repeated this same testimony to the Senate Subcommittee on Environmental Pollution. What Perhac did not tell either the representatives or the senators is that the three lakes in question—and they happen to be Panther, Sagamore, and Woods, Bill Marleau's old favorite—have different buffering capacities.

• *Sudden acidification of a body of water, in itself, may not be responsible for fish kills.*

Perhac used this argument in his testimony before both the House and the Senate subcommittees in 1980, and he cited the case of a fish kill that occurred in the Tovdal River in Norway in early 1975. It is correct that the sudden acidification "in itself," to quote Perhac's hedge phrase, did not kill the fish in the Tovdal River. What Perhac did not say was that Norwegian scientists determined that the likely killing agent was aluminum mobilized by the melt of acid snow.

• *Acid rain is turning up in remote parts of the world, such as Hawaii. Therefore acid rain is natural and industry cannot be blamed.*

This argument is completely irrelevant to the situation in the northeastern United States and Canada, where natural sources are far too small to account for the observed sulfuric acid in precipitation. Yes, rain in Hawaii is acid, ranging from pH 5.2 at sea level to 4.3 at 7,500 feet, but the scientists who documented this, John M. Miller and Alan Yoshinaga of the National Oceanic and Atmospheric Administration, have suggested that convective rainstorms may reach high up into the troposphere to precipitate pollutants coming from distant sources. Canadian scientists recently reported that pollutants traveling 3,000 miles and more from Europe, Asia, and possibly North America are causing a pervasive haze in the Arctic during winter and spring. As Miller wrote in a paper published in *Acid Rain*, the proceedings of a session sponsored in 1979 by the American Society of Civil Engineers, ". . . acid rain is a problem, not only of national but of global dimensions. . . ."

• *A reduction in sulfur-dioxide emissions in the Midwest and Northeast, say of 50 percent, would not cause a 50 percent decline in suspended sulfates, wet sulfur deposition, and acid precipitation in the Northeast.*

"It is true that there is not a one-to-one correspondence in reductions," said Michael Oppenheimer, of the Environmental Defense Fund, who has been studying the chemical transformation and deposition of sulfur, "but long-range transport models indicate that fifty percent reductions would lead to very significant decreases closer to fifty percent than to zero."

• *Any acidified waters could be restored by liming.*

Liming is expensive, and it is useful only on a limited Band-Aid scale for the preservation of unique fish populations. "It has its place currently in fishery management," said Dr. Carl Schofield, of Cornell University, "but it isn't viewed as the solution to the problem." Dr. Harold Harvey of the University of Toronto said, "Let us dismiss out of hand that we can lime the northeast quadrant of a continent.'' Moreover, Harvey pointed out, ''If you take an acid lake and lime it, you do not now have a normal lake; you now have a limed, formerly very acid lake, with a very peculiar water chemistry and a very peculiar biota as a result.'' Liming also does not answer the other threats posed by acid precipitation, such as the deposition of metals.

• *''In 1980 two scientists who tested ice core samples concluded that acid rain existed long before the Industrial Revolution. They found the samples, which were taken from the Antarctic and Himalayan Mountains, laden with acid. One sample was 350 years old.''*

This ''well-documented and proven information'' is cited in the Edison Electric Institute's publication, *Before the Rainbow: What We Know About Acid Rain*. This information is false. It was based on an article that ran in *The Wall Street Journal* on September 18, 1980. That story began, ''Acid rain, a recent concern of environmentalists, has been pelting the earth for centuries, according to findings by two University of New Hampshire scientists.'' The two scientists referred to, Dr. Paul Mayewski and Dr. W. Berry Lyons (erroneously called Barry in the story), insist they made no such findings. ''The story was extremely distorted,'' said Mayewski. ''There were no significant heavy-acid traces at all in the cores, and we stressed to the man from *The Wall Street Journal* that we were doing research on ice and snow, not rain and acid rain.'' Asked about this, Mitchell C. Lynch, the reporter whose byline appears on *The Wall Street Journal* story, stood by the article as an accurate presentation of the information given him by the scientists. On October 1, 1980, the *Journal* used Lynch's article as a peg for an editorial declaring that the "theory" that acid rain is a result of industrialization "has just taken a couple of body blows from Mother Nature."

Others have taken up this cry. In the March–April 1981 issue of *Bassmaster Magazine*, Jene L. Robinson, director of Environmental Affairs for the Illinois Power Company, wrote, "Scientists at the University of New Hampshire have found 350-year-old ice samples in the Himalayas and Antarctica that have pH in the 4.5-to-5.1 range. Thus, it may well be that 'normal' rain would be more acidic than 'acid' rain." In a speech in May 1981, at an international conference on acid rain at the State University of New York at Buffalo, William N. Poundstone, executive vice-president of the Consolidation Coal Company, cited "research by two scientists at the University of New Hampshire. They studied Antarctic and Himalayan ice cores, dating back 350 years. Here clearly in the absence of fossil-fuel-burning plants, they found pH values in the low 5s."

If false information is repeated often enough, it can creep into respected sources. Thus the *Congressional Research Service Review*, published by the Library of Congress, noted in an article on acid precipitation that "Industry and other skeptics, however, believe that present knowledge of the acid-rain problem is insufficient for regulatory action. For example, the extent to which coal burning contributes to acid rain is questioned by the coal and utility industries. They cite results of tests made by two University of New Hampshire scientists who measured the acidity of ice core samples from Antarctica and the Himalayan Mountains. The samples were 'laden with acid.' "

Take the phrase "rush to judgment," or a "pell-mell rush to judgment," or just the verb "rush." Industry apologists use "rush" anytime anyone proposes action on acid precipitation. In February 1980, William B. Harrison, senior vice-president of the Southern Company Services, speaking on behalf of the Utility Air Regulatory Group and the Edison Electric Institute, told the House Subcommittee on Oversight and Investigations of the Committee on Interstate and Foreign Commerce that "In our judgment, it is unwise and unnecessary for this committee to rush to a premature legislative solution to the acid-precipitation problem before the problem can be clearly defined and understood." That June, associate editor R. C. Rittenhouse warned in *Power Engineering*, "Pressures have been mounting to rush into legislation . . ." and later in his column he noted that William N. Poundstone of Consolidation Coal had "observed this same rush to regulate." Indeed Poundstone had. In 1981 Poundstone told the *Cleveland Plain Dealer*, "The worst thing we can do is to rush headlong into a regulatory plan." A. Joseph Dowd, senior vice-president and general counsel of American Electric Power, agreed. He called his May 1981 speech at the international acid-rain conference in Buffalo, "Costs Without Benefits, The Rush to Judgment on Acid Rain," and in it he cited the risks inherent in "rushing to judgment on the acid-rain issue," and he then rushed to add "that there is no need to rush to judgment." The Coalition for Environmental-Energy Balance, a front for Midwestern utilities and manufacturers, rushed into print with a national newspaper ad campaign attacking legislation under consideration before the Senate Committee on Environment and Public Works "as a rush to judgment, if you will," and the coalition rushed publication of a brochure that denounced "those who rush to judgment with a quick-fix legislation," asked "What's the big rush?", pledged to "prevent a rush to judgment," and urged readers to write to members of the Senate committee to protest against "a 'quick-fix' rush to judgment."

A point to bear in mind: Utility companies love brochures, pamphlets, or articles planted in house organs that they try to palm off as scientific reports or analyses. "Utilities don't publish papers in scientific journals with peer review," Fred Johnson of the Pennsylvania Fish Commission said. "They put out propaganda, and this confuses the public, which is going to suffer in the end. I just got another piece of garbage in the mail today from General Public Utilities." The GPU brochure that so exercised Johnson is called *Take the Acid Test*, and one of the questions it asks is "Is acid rain a problem in Pennsylvania?" The GPU's answer is, "The results of studies to date are inconclusive and often downright contradictory. For example, the Pennsylvania Fish Commission is blaming increasing acidity in the rainfall for low fish populations in some streams. But Dr. Robert P. Pfeifer, an associate professor at Pennsylvania State University, views acid rain as a boon to the Pennsylvania farming community. He said that without the sulfur and nitrogen brought down by acid rain, Pennsylvania would become barren of most vegetation."

Inasmuch as industry spokesmen all seem to be reaching into the same old bag, it is sometimes difficult to determine who wrote what first. Two articles published back to back in *Before the Rainbow: What We Know About Acid Rain*, a recent 102-page paperback that the Edison Electric Institute published as part of its—and we invent nothing—"Decisionmakers Bookshelf," are just about identical, word for word, page after page, except for the placement of some paragraphs.

Here is a sample paragraph from the first of the two articles, "Acid Precipitation: The Issue in Perspective," no date or author given, but billed in the book as the "industry perspective":

The chemistry of precipitation was first studied for its nutrient qualities. Nitrogen and chlorine compounds in rain were first examined in the early 1800s, and sulfur compounds in the early 1900s. However, the pH of precipitation did not become an issue of concern until the 1950s,

Here for comparison is a paragraph from the article immediately following, "Acid Precipitation—A Review of the Issue and Its Uncertainties," by John J. Jansen, a research specialist for Southern Company Services:

The chemistry of precipitation was first studied for its nutrient qualities. Nitrogen and chlorine compounds were first studied in the early 1800's and sulfur compounds in the early 1900's. The subject of the acidity of precipitation did not become an issue of concern until the 1950's, when the when the Scandinavian countries began monitoring precipitation chemistry. The increased acidification of lakes in Scandinavia and the disappearance of fish populations, first reported in the 1920s, were attributed to a purported increase in precipitation acidity. As a result, the chemistry of precipitation, once generally thought to be beneficial, is now viewed by many as harmful.

Scandinavian countries began monitoring precipitation chemistry. The increased acidification of lakes and the disappearance of fish populations, first reported in the 1920's, were soon attributed to the purported increase in precipitation acidity. As a result, the chemistry of precipitation turned from being viewed as beneficial to being harmful. . . .

Here is another extract from the first article:

1. *Historical Data and Trends*

There have been numerous claims that acidity in precipitation has been increasing in both magnitude and area for the past 25 years. The most often cited evidence for this contention is a paper written by C. V. Cogbill and G. E. Likens in 1974.

Essentially, Cogbill and Likens used precipitation chemistry data from the 1950s and 1960s to calculate the pH of the precipitation in those years. Then, by plotting the values on a map of the eastern United States drawing iso-

From the second article:

Historical data and trends

A prevalent opinion is that acid precipitation has been increasing in both magnitude and area for the past 25 years. The most often cited evidence for this contention is a paper written by Cogbill and Likens in 1974.

Essentially, Cogbill used precipitation chemistry data from the 1950's and 1960's to calculate what the pH of the precipitation was in those years. Then by plotting the values on a map of the eastern United States drawing isopleths of equal pH, and com-

pleths of equal pH, and comparing maps for succeeding time periods, they concluded that the acidity in precipitation was increasing in magnitude in the Northeast and spreading west and south. In addition, they attributed this increase in sulfur and nitrogen oxide emissions and the increase in the use of tall stacks by power plants and other industrial sources.

paring maps for succeeding time periods, he concluded that the acidity in precipitation was increasing in magnitude in the Northeast and spreading towards the West and South. In addition, he went on to attribute this increase in acidity to the concurrent increase in sulfur and nitrogen oxide emissions and the increase in the use of tall stacks by power plants and other industrial sources.

And so the twin articles go on, paragraph after paragraph. Asked who wrote the first article, Carolyn Curtis, the editor of *Before the Rainbow,* replied, "I'm not sure of the individual who wrote that. I thought it was the weakest of the ones given to me." Where does that put the second article? And where does that put the book when these two very similar articles make up almost half of it? One of the other articles in the book, *mirabile dictu,* is Dr. Ralph ("Three Lakes in the Adirondacks") Perhac's testimony to the Senate Subcommittee on Environmental Pollution on March 19, 1980. Industry apologists are not only reaching into the same old bag but they are going through revolving doors at the same time.

It may seem difficult to believe, but *Before the Rainbow* was specifically compiled to define the stance of the utility industry on acid rain. As Curtis told us, it is intended to influence "decision makers, committee staffers, CEO's, industry members who want to repeat the party line." In her introductory chapter, Editor Curtis sets the tone with an approach that reads like Dick and Jane meet acid rain. After seizing upon the fact that "natural rain is somewhat acidic," Curtis wrote, "So by all rights we should have been saying for years, 'It's acid raining outside,' or 'Take your umbrella. It's going to acid rain today.' This sounds preposterous, but it's true. Thus, our first understanding is that the strong verbal image, 'acid rain,' elicits more fear than it deserves."

Curtis, a Washington consultant whose experience includes "all phases of corporate and industrial communication" and who has won, the book points out, awards from the International Association of Business Communicators, Society for Technical Communication, and Society of Professional Journalists, also wrote in *Before the Rainbow* that

> what has been printed on this subject [of acid rain] ranges from good through mediocre to bad in terms of editorial consistency and scientific soundness. Results: enlightenment, interest, concern, misunderstanding, confusion. An emotion-charged issue, about which not enough is truly known and proven, usually results in that spectrum.
>
> The media—which have many fine writers, editors and, yes, thinkers—have painted for me an amusing portrait of two of our society's most distinguished professions. One is of a scientist who one day comes across some surprising information: that precipitation is higher in acidity than distilled water.

He scratches his head and says, "By golly, I've been wondering why all those fish have been disappearing!" The other is of a government lawmaker. He reads a report that scientists are learning rain has a higher acid content than they realized before. He looks up at a staffer and shouts, "I knew it! It's those so-and-sos in industry. They've been sending stuff up in the air and now it's showering down on all of us."

If scientists and Congressmen were as overly reactive and quick to jump to conclusions as that, we would not have progressed beyond the alchemists and feudal system of the Middle Ages. On the other hand, if we knew all the scientific information there was to know and if every law had been passed, then those folks wouldn't have much to do.

7

The Solution

Acid precipitation is a very real and very frightening problem. As we noted in Chapter 1, a committee appointed by the National Academy of Sciences stated that "the picture is disturbing enough to merit prompt tightening of restrictions on atmospheric emissions." Other groups, including the National Commission on Air Quality, the National Governors Association, and the Organization of State and Local Air Pollution Officials, agree. The National Clean Air Coalition, which is composed of the American Lung Association, the League of Women Voters, the United Steelworkers of America, the Natural Resources Defense Council, the Environmental Defense Fund, the International Association of Machinists, the National Wildlife Federation, the National Audubon Society, Friends of the Earth, the Sierra Club, the Izaak Walton League of America, and the National Parks and Conservation Association, is endeavoring to put an end to acid precipitation.

The ways and means to end acid precipitation are not arcane. They are readily at hand, and they should be made part of a stronger Clean Air Act. The revised law should require emission reductions that can be accomplished by the establishment of emission standards on a regional or "bubble" basis, the burning of low-sulfur coal, the installation of scrubbers at critical plants, and investment in alternative energy sources. The costs would be very small compared to the rate hikes imposed by OPEC in recent years. The National Commission on Air Quality, whose members were appointed by President Jimmy Carter, reported that a two percent surcharge on the average utility bill in the East would bring in enough money to eliminate half the sulfur dioxide in the region.

There are other benefits, economic benefits, according to Russell W. Peterson, the president of the National Audubon Society, who has calculated that the measurable benefits of air-pollution controls would outweigh costs by approximately $5 billion a year. Manufacturing scrubbers and other clean-up devices could provide both jobs and opportunities—and clean air.

In October of 1981, Louis Harris, the pollster, appeared before the House Subcommittee on Health and the Environment of the Energy and Commerce Committee, and he testified that "by 80 to 17 percent, a sizable majority of the public does not want to see any relaxation in existing federal regulation of air pollution." This was true of every major segment of the public, including those who had voted for Reagan in 1980, and Harris testified that "the mandate for Ronald Reagan's election did not include in any shape or form a repeal or watering down of the tough provisions" of the Clean Air Act. Harris concluded, "The American people *are* willing to make sacrifices in many areas to stop the miseries and ravages of inflation and an economy that is out of joint. But they will not tolerate any reductions in environmental clean-up efforts—and will regard such cuts as threatening to the very quality of life in this last quarter of the twentieth century. I am not an expert on this legislation nor on the subject of environmental regulation, but I can tell you this: this message on the deep desire on the part of the American people to battle pollution is one of the most overwhelming and clearest we have ever recorded in our 25 years of surveying public opinion."

The Reagan Administration is buying none of this. It is out to gut the Clean Air Act (and the Clean Water Act), and it has allies in Congress, notably Rep. Thomas Luken, Democrat of Ohio; Rep. John Dingell, Democrat from Michigan; and Rep. James Broyhill, Republican from North Carolina, who at this writing are cosponsoring a "Dirty Air Act" that would roll back clean-air standards for new cars, weaken provisions governing pollution standards for trucks, weaken programs designed to make certain that auto-pollution-control equipment works, and ignore the problem of acid precipitation.

Richard Ayres, chairman of the National Clean Air Coalition, has been a firsthand observer of the administration's maneuvering on the Clean Air Act, and he stated,

"As much as anything I've ever seen in Washington, this shows that organized business can thwart the interests of the people." The very day that Ayres made this remark to us, Energy Secretary James B. Edwards showed up in Chicago to address the American Power Conference, and when he was through Edwards said, "I think we need to do more research before we run off and worry about acid rain falling from the sky. I don't want to stop acid rain because the fields are alkaline, and a little acid rain helps neutralize the soil." Edwards added, "If you dig down in the glacial ice, it's more acid than the rain we have today, so I wonder what smokestacks from a couple of billion years ago were responsible."

Recently *The Clarion-Ledger* of Jackson, Mississippi, which had become concerned that increased acidity in Mississippi soils could cause short-term surges of toxic aluminum, reported that Anne Gorsuch, administrator of the Environmental Protection Agency, had slowed the flow of research information to the public by devising unwritten regulations governing publication. The paper quoted an EPA official, who asked not to be identified, as saying, "Our new peer review policy for releasing research from EPA is a joke. We can have up to seven tiers of review before release. If you flunk any one of the seven, you go back to the beginning. This is true for talks, computer programs or anything you want to talk about. These rules have not been issued in written form by the administrator's office, but we're supposed to act as though they have.

"We have an incredible double standard working in

the agency now," the official continued. "For those areas where the administration already knows what it wants to do, it requires no scientific proof. It goes out and looks for a preponderance of evidence. But if the preponderance of evidence indicates that regulations are required where the administration does not want to apply them, scientists are asked to come up with absolute proof of their findings. So we have a different scientific standard for what the administration wants and what it doesn't want.

"That, to me, smacks of being dishonest," the official said. "They're hiding behind this great science banner which they are using in a prostituted manner. I'm ashamed that politicians have become book-burners and censors. The public has paid for this information, and they will not let the public see it."

At this writing, there is growing public concern about the proliferation of nuclear weapons and the threat of global annihilation. Without attempting to minimize this concern, we frankly wonder why there is all the fuss about nuclear weapons. We don't need the Soviet Union to destroy us. Given the irresponsibility of the Reagan Administration, this country is quite capable of destroying itself in pursuit of bizarre economic ideology. Given our present political course and the devastation that acid rain, acid precipitation, or acid deposition can bring, T. S. Eliot was right when he wrote in "The Hollow Men,"

This is the way the world ends,
This is the way the world ends,
This is the way the world ends,
Not with a bang but a whimper.

Appendix 1

In 1981 the National Wildlife Federation, using data from published studies, precipitation-chemistry monitoring stations, and personal communications from scientists and others, assessed the "acid-rain vulnerability" of the twenty-seven states east of the Mississippi River based on the potential damage to fisheries, soils, crop foliage, and masonry. The report also considered visibility impairment and automobile paint damage based on existing rather than potential impact.

A number of western states are known to be sensitive to acid precipitation, but the National Wildlife Federation limited the report to the eastern states because of the availability of data. We have some minor differences with the report—we are of the opinion, for example, that Florida is more vulnerable than indicated—but the report demonstrates the pervasiveness and potential severity of the problem in the eastern states.

The table prepared for the report follows.

EVALUATION OF EASTERN STATES' VULNERABILITY TO EFFECTS OF ACID RAIN

State	pH RANGE OF RAINFALL	AVERAGE pH OF RAINFALL	FISHERIES	SOIL	CROP FOLIAGE	MASONRY (MARBLE, LIMESTONE)	OVERALL RANKING	VISIBILITY IMPAIRMENT	AUTOMOBILE PAINT DAMAGE
Alabama	3.7–. . .	4.6	M	M	V	M	M	E	S
Connecticut	3.6–. . .	4.4	E	E	V	M	E	E	S
Delaware	3.5–5.0	4.4	M	M	V	M	M	E	M
Florida	4.2–6.1	4.7	S	S	NV	S	S	M	E
Georgia	4.1–6.3	4.6	M	M	NV	S	M	M	S
Illinois	3.8–6.7	4.3	S	S	V	M	M	E	M
Indiana	3.8–. . .	4.3	S	M	V	M	M	E	M
Kentucky	3.7–. . .	4.4	M	E	V	M	E	E	M
Louisiana	XX	4.6	XX	M	XX	S	XX	E	S
Maine	3.4–6.9	4.4	E	E	V	M	E	S	S
Maryland	3.7–. . .	4.3	M	M	V	M	M	E	S
Massachusetts	3.5–5.7	4.1	E	E	V	M	E	E	M
Michigan	3.8–6.1	4.4	E	E	V	M	E	E	M
Mississippi	4.2–5.3	4.6	M	M	NV	S	M	E	S
New Hampshire	3.7–6.2	4.3	E	E	V	M	E	S	M
New Jersey	3.5–. . .	4.3	E	M	V	M	E	E	E
New York	3.4–5.1	4.2	E	E	V	M	E	E	E
North Carolina	3.3–6.9	4.4	E	E	V	M	E	M	M
Ohio	3.6–5.1	4.2	S	M	V	M	M	E	M
Pennsylvania	3.5–5.9	4.2	E	E	V	M	E	E	E
Rhode Island	3.5–. . .	4.4	E	E	V	M	E	E	M
South Carolina	3.2–4.7	4.6	E	M	V	M	E	M	S
Tennessee	3.5–. . .	4.4	M	M	V	M	M	E	S
Vermont	3.5–. . .	4.3	E	E	V	M	E	M	M
Virginia	3.1–5.1	4.4	M	M	V	M	M	E	M
West Virginia	3.8–6.3	4.3	M	E	V	M	E	E	M
Wisconsin	3.8–6.1	4.5	E	E	V	M	E	E	M

E—Extremely vulnerable to acid-rain effects
M—Moderately vulnerable to acid-rain effects
S—Slightly vulnerable to acid-rain effects
V—Vulnerable to acid-rain effects
NV—Not vulnerable to acid-rain effects
XX—Insufficient data

Appendix 2

Organizations involved in efforts to curb acid deposition include the following:

American League of Anglers
810 18th Street, N.W.
Washington, D.C. 20006

Canadian Coalition on Acid Rain
1825 K Street, N.W.
Washington, D.C. 20006

Canadian Wildlife Federation
Suite 106
1673 Carling Avenue
Ottawa, Ontario K2A 1C4, Canada

Environmental Defense Fund
444 Park Avenue South
New York, NY 10016

Friends of the Earth
1045 Sansome Street
San Francisco, CA 94111

Izaak Walton League of America
1800 North Kent Street
Arlington, VA 22209

National Audubon Society
950 Third Avenue
New York, NY 10022

National Clean Air Coalition
530 7th Street, S.E.
Washington, D.C. 20003

Natural Resources Defense Council
122 East 42nd Street
New York, NY 10168

National Wildlife Federation
1412 16th Street, N.W.
Washington, D.C. 20036

Sierra Club
530 Bush Street
San Francisco, CA 94108

Trout Unlimited
118 Park Street
Vienna, VA 22180

Information on acid precipitation and materials for schools and colleges can be obtained from:

The Acid Rain Foundation, Inc.
1630 Blackhawk Hills
St. Paul, Minnesota 55122

Bibliography

We have drawn on a number of references dealing with the subject of acid precipitation in general. For the sake of convenience, we have listed them here chronologically.

Harold H. Izard and Jay S. Jacobson, editors, *Scientific Papers from the Public Meeting on Acid Precipitation, May 4-5, Lake Placid, New York* (mimeographed and published by Science and Technology Staff, New York State Assembly, Albany; March, 1979). Papers of interest are Raymond Falconer, "Acid Rain and Precipitation Chemistry at Whiteface Mt., N.Y."; Carl L. Schofield, "The Acid Precipitation Phenomenon and its Impact in the Adirondack Mountains of New York State," which deals with Schofield's finding of the toxicity of aluminum during the spring snowmelt; George R. Hendrey and Frank W. Barvenik, "Impacts of Acid Precipitation on Decomposition and Plant Communities in Lakes"; and Christopher S. Cronan, "Effects of Acid Precipitation on Soil Leaching Processes in High Elevation Coniferous Forests of the Northeastern U.S."

Proceedings of the Action Seminar on Acid Precipitation, Nov. 1st to 3rd, 1979 (no editor, no date and no place of publication, but presumably Toronto where the ASAP [Action Seminar on Acid Precipitation] Coordinating

Committee held the conference). Papers of note include Ellis B. Cowling, "From Research to Public Policy: Progress in Scientific and Public Understanding of Acid Precipitation and its Biological Effects"; David W. Schindler, "Effects of Acid Deposition on Canadian Lakes and Fisheries"; Eville Gorham, "A Recommended Program of Research"; George R. Hendrey, "Acidification of Aquatic Ecosystems: Ecosystem Sensitivity and Biological Consequences"; David B. Peakall, "Acid Precipitation and Wildlife"; L. D. Hamilton, "Health Effects of Acid Precipitation"; P. J. Rennie, "Dangers to Soils and Vegetation"; and Gus Speth, "The Sisyphus Syndrome: Air Pollution, Acid Rain and Public Responsibility."

Acid Rain, Hearings before the Subcommittee on Oversight and Investigations of the Committee on Interstate and Foreign Commerce, House of Representatives, Ninety-sixth Congress, Second Session, February 27 and 28, 1980 (Washington, 1980) contains statements, from which we have drawn, by Ellis Cowling, Eville Gorham, Walter A. Lyons of Mesomet, and Ralph Perhac of the Electric Power Research Institute.

The Norwegian SNSF Project published two works of importance. The first is D. Drabløs and A. Tollan, editors, *Ecological Impacts of Acid Precipitation* (Oslo, 1980), the proceedings of an international conference held at Sandefjord, Norway, March 11-14, 1980. This is a very important reference, a key work in the field. We have no wish to compile a bibliography of lists, but here are papers we consulted: J. S. Jacobson, "The Influence of Rainfall Composition on the Yield and Quality of Agricultural

Crops"; A. Henriksen, "Acidification of Freshwaters—a Large Scale Titration"; I. P. Muniz and H. Leivestad, "Acidification—Effects on Freshwater Fish"; H. H. Harvey, "Widespread and Diverse Changes in the Biota of North American Lakes and Rivers Coincident with Acidification"; S. J. Eisenreich and colleagues, "Impact of Land-Use on the Chemical Composition of Rain and Snow in Northern Minnesota"; G. E. Glass, "Susceptibility of Aquatic and Terrestrial Resources of Minnesota, Wisconsin and Michigan to Impacts from Acid Precipitation: Informational Requirements"; G. E. Glass and colleagues, "The Sensitivity of the United States' Environment to Acid Precipitation"; C. D. Hendry and colleagues, "Acid Precipitation in Florida (U.S.A.): Results of a Statewide Monitoring Network"; J. G. McColl, "Acid Precipitation and Ecological Effects in Northern California"; R. Herrmann and J. Baron, "Aluminum Mobilization in Acid Stream Environments, Great Smoky Mountains National Park, U.S.A."; H. G. Jones and colleagues, "The Evolution of Acidity in Surface Waters of Laurentides Park (Quebec, Canada) over a Period of 40 Years"; M. E. Thompson and colleagues, "Evidence of Acidification of Rivers in Eastern Canada"; J. H. D. Vangenechten and O. L. J. Vanderborght, "Acidification of Belgian Moorland Pools by Acid Sulphur-Rich Rainwater";

R. F. Wright and colleagues, "Acid Lakes and Streams in the Galloway Area, Southwestern Scotland"; T. L. Chrisman and colleagues, "Acid Precipitation: the Biotic Response in Florida Lakes"; H. Leivestad and colleagues, "Acid Stress in Trout from a Dilute Mountain

Stream"; K. Tonnessen and J. Harte, "The Potential for Acid Precipitation Damage to Aquatic Ecosystems of the Sierra Nevada, California (U.S.A.)"; I. H. Sevaldrud and colleagues, "Loss of Fish Populations in Southern Norway. Dynamics and Magnitude of the Problem"; and R. F. Wright and colleagues, "Acidified Lake Districts of the World: a Comparison of Water Chemistry of Lakes in Southern Norway, Southern Sweden, Southwestern Scotland, the Adirondack Mountains of New York, and Southeastern Ontario."

The second SNSF publication is Lars N. Overrein, Hans Martin Seip, and Arne Tollan, editors, *Acid Precipitation—Effects on Forest and Fish* (Oslo; December, 1980), which is the final report of the 1972–1980 SNSF project. It is based on almost 300 reports and papers published in journals throughout the world, and although it concentrates on Norway, the findings are of relevance to other areas afflicted by acid precipitation.

Two books that deal mainly with the Canadian experience and political wrangling within Canada are Ross Howard and Michael Perley, *Acid Rain, The North American Forecast* (House of Anansi Press Limited, Toronto; 1980), and Phil Weller and the Waterloo Public Interest Research Group, *Acid Rain, the Silent Crisis* (Between The Lines, Kitchener, Ontario; 1980).

We have read with interest the following reports and working papers of the "United States–Canada Memorandum of Intent on Transboundary Air Pollution." We are aware, of course, that these are not final reports, so wherever possible we have attempted to confirm the information reported by interviewing the participants and their peers in the field. *Interim Report, February 1981*, from which we have taken in Section 2, Table 2-1, the summary of global emissions of sulfur dioxide, nitrogen oxides, and hydrocarbons given in Chapter 1; *Impact Assessment, Interim Report, February 1981; Strategies Development and Implementation, Interim Report 1981;* and *Phase II Interim Working Paper, October 1981*.

A very important work is that published by the National Academy Press of the National Academy of Sciences. It is a report prepared by the committee on the Atmosphere and the Biosphere of the National Research Council, *Atmosphere-Biosphere Interactions: Toward a Better Understanding of the Ecological Consequences of Fossil Fuel Combustion* (Washington, D.C., 1981). Although of necessity technical, it is very readable, and best of all it is indexed. Of special interest to us—and the entire report was essential reading—was the introductory and overview chapter and the chapters on anthropogenic (manmade) sources of atmospheric substances, the biological accumulation and effect of atmospheric contaminants, studying the effects of atmospheric deposition on ecosystems, and acid precipitation. In sum, a documented basic study.

The most concise paper, a review complete with references, that we have read is by Lance S. Evans, George R. Hendrey, and colleagues, "Acidic Precipitation: Considerations for an Air Quality Standard," delivered at the Air Pollution Control Association Specialty Meeting, Atlanta, Georgia, September 16-18, 1980, and scheduled for publication in *Water, Air, and Soil Pollution*, Vol. 16, 1981.

The National Wildlife Federation and the Environmental Defense Fund published a very useful summary, *Acid Precipitation: A Problem in Need of Immediate Legislative Remedy* (Washington, 1981). At the same time, the National Wildlife Federation also issued a report, "Acid Rain Vulnerability of the 27 States East of the Mississippi River," from which we have taken the table in Appendix 1 by permission.

On a regional basis, we have consulted the following works: Martin H. Pfeiffer and Patrick J. Festa, "Acidity Status of Lakes in the Adirondack Region of New York in Relation to Fish Resources" (New York State Department of Environmental Conservation, August 1980); Jennie E. Bridge and F. Peter Fairchild, *Northeast Damage Report of the Long Range Transport & Deposition of Air Pollutants* (Northeast States for Coordinated Air Use Management, Weston, Massachusetts, and the New England Interstate Water Pollution Control Commission, Boston; April 1981), from which we have quoted extensively; and *Acid Rain, Report to The Governor* [of Ohio] *by the Scientific Advisory Task Force* (Ohio Agricultural Research and Development Center, Wooster, Ohio; September, 1981). For Ohio's attitude toward acid precipitation, we have also used a series of articles by Richard G. Zimmerman and Amos A. Kermisch, "Poison from the Skies?" reprinted from *The Plain Dealer*, Cleveland, August 2-7, 1981. In August of 1981, we attended an International Symposium on Acidic Precipitation and Fishery Impacts in Northeastern North America at Cornell University. Presented by the Northeastern Division of the American Fisheries Society and sponsored by the U.S. Fish and Wildlife Service and the Canadian Department of Fisheries and Oceans, the symposium attracted authorities from the U.S., Canada, and Europe. Unfortunately, the American Fisheries Society had yet to publish the papers presented at the symposium as this book went to press. Previously we attended the Conference on Acid Rain and the Atlantic Salmon in Portland, Maine, in November, 1980, and the proceedings, *Acid Rain and the Atlantic Salmon*, were published in March, 1981, by The International Atlantic Salmon Foundation in New York City and St. Andrews, New Brunswick. Papers of note are Harold H. Harvey, "Where Have All the Fishes Gone?" Carl L. Schofield, "Aquatic Effects of Acid Rain"; Peter G. Daye, "The Impact of Acid Precipitation on the Physiology and Toxicology of Fish"; W. D. Watt, "Present and Potential Effects of Acid Precipitation on the Atlantic Salmon in Eastern Canada"; and Terry A. Haines, "Effects of Acid Rain on Atlantic Salmon Rivers and Restoration Efforts in the United States."

The sensitivity of rocks and soils to acid precipitation is of basic importance, and works we have used are William W. McFee, *Sensitivity of Soil Regions to Acid Precipitation* (U.S. Environmental Protection Agency, Corvallis, Oregon; EPA-600/3-80-013; January, 1980); George R. Hendrey, James N. Galloway, and colleagues, *Geological and Hydrochemical Sensitivity of the Eastern United States to Acid Precipitation* (U.S. EPA, Corvallis, Oregon; EPA-600/3-80-024; January, 1980); and Stephen A. Norton, Ronald B.

Davis, and colleagues, *Responses of Northern New England Lakes to Atmospheric Inputs of Acids and Heavy Metals* (University of Maine at Orono; Project A-048-ME; July, 1981); and W. W. Shilts, *Sensitivity of Bedrock to Acid Precipitation: Modification by Glacial Processes* (Geological Survey of Canada, Ottawa; 1981), which is accompanied by three maps of south-central and southeastern Canada.

For vegetation, we have used Bruce Haines, Marcia Stefani, and Floyd Hendrix, "Acid Rain: Threshold of Leaf Damage in Eight Plant Species from a Southern Appalachian Forest Succession," *Water, Air, and Soil Pollution,* Vol. 14 (Boston, 1980); three articles in *Der Spiegel,* the West German news magazine, beginning with "Saurer Regen über Deutschland," in the November 16, 1981 issue and including the November 23 and November 30 issues; G. H. Tomlinson, "Acid Rain and the Forest—The Effect of Aluminum and the German Experience" (Domtar Research Centre, Senneville, Quebec; February, 1981) and "Die-Back of Forests—Continuing Observations—May and June 1981" (Domtar Research Centre, Senneville, Quebec; June, 1981); and F. H. Bormann's manuscript draft, "The New England Landscape: Air Pollution Stress and Energy Policy," part of which was published in the *Yale Alumni Magazine and Journal,* April, 1982, before publication in Carl Reidel, editor, *New England Prospects: Critical Choice in a Time of Change* (University Press of New England, Hanover, New Hampshire; 1982).

For the influence of air pollution on weather and climate, we used *Effects of Chronic Exposure to Low-Level Pollutants in the Environment,* prepared for the Subcommittee on the Environment and the Atmosphere of the Committee on Science and Technology, U.S. House of Representatives, Ninety-fourth Congress, by the Congressional Research Service, Library of Congress, Serial O (Washington, November, 1975), which deals with the La Porte Weather Anomaly and other effects on climate and weather on pages 265-299; Stanley A. Changnon, Floyd A. Huff, and colleagues, *Summary of METROMEX, Volume 1: Weather Anomalies and Impacts* and *Summary of METROMEX, Volume 2: Causes of Precipitation Anomalies* (Illinois State Water Survey, Urbana; 1977, 78); and Council on Environmental Quality, *Global Energy Futures and the Carbon Dioxide Problem* (Washington; January, 1981).

Index